ELECTRICITY
A STEP-BY-STEP GUIDE

SMITHMARK

© 1993 Dragon's World Ltd

Photographs by Jon Bouchier, Simon Butcher, Simon Wheeler.

Illustrations by Kuo Kang Chen, Steve Cross, Paul Emra, Pavel Kostal, Janos Marffy, Sebastian Quigley, Laurie Taylor, Brian Watson, Andrew Green.

This edition published in 1993 by SMITHMARK Publishers Inc., 16 East 32nd Street, New York, NY 10016.

SMITHMARK books are available for bulk purchase for sales promotion and premium use. For details write or call the manager of special sales, SMITHMARK Publishers Inc., 16 East 32nd Street, New York, NY 10016; (212) 532-6600.

Produced by Dragon's World Ltd, 26 Warwick Way, London SW1V 1RX, England.

Editor: Dorothea Hall
Designers: Bob Burroughs, Mel Raymond
Art Director: Dave Allen
Editorial Director: Pippa Rubinstein

ISBN 0-8317-4626-2

Printed in Italy

10 9 8 7 6 5 4 3 2 1

CONTENTS

INTRODUCTION

Providing comfort and convenience, electricity helps to keep our homes warm and our refrigerators cool — two of the many facilities we would find difficult to live without. Fortunately for the home-improver, electrical work is one of the easiest kinds of home maintenance and repair. You do not require a great variety of specialized tools, and to make things easier, many home electrical systems and related materials are now standardized. However, before you undertake any wiring projects, it is extremely important to understand how electricity works in your home.

The following pages explain this along with information on how to do simple repairs, add lighting and power circuits, install appliances and take lighting outdoors, so that you can safely and confidently use electricity in your home with maximum effect.

However competent the average householder may be in other areas of DIY, some people are extremely reluctant to attempt all but the simplest of electrical work in the home. Since electricity can neither be smelled nor seen, this attitude is, to a certain extent, understandable. And with something as potentially dangerous as electricity, it would be foolhardy for anyone to attempt a major installation without first working on simpler tasks to familiarize themselves with what is involved.

Safety first

In practice the skills involved in do-it-yourself electrical work are easy to master, and furthermore you can always see what you are doing and double-check that you have done it properly ... unlike plumbing, for example, where you can only tell if your joints are watertight when you turn the water back on. So provided that you allow sufficient time (never take short cuts and try to hurry a task), and that you are able to work methodically and are prepared to check everything twice, there is no reason to be afraid that the end result will be at all dangerous. You will probably have taken a great deal more care than the average electrician, simply because it is your own home and your family that are involved.

The other quality you need is respect: electricity can be a wonderfully willing servant so long as you treat it properly. At all times, take sensible precautions when you are carrying out electrical work, and you will be at no risk. This means never attempting any electrical job unless you know exactly what you are doing. Never working on any part of your electrical system unless you have turned off the branch circuit on which you are working, and always double-checking everything thoroughly before restoring the power. However, should you begin and then doubt your ability, do get professional help.

Last of all, teach your children about the dangers of electricity, so that they will understand it and not take chances with it when they grow up.

Local codes

All electrical work whether done by an electrician or not must be carried out according to your local building code and the National Electrical Code (NEC), which is available at your local library. The codes may vary greatly from one location to the next, so always check the proposed work with your local Building Department to make sure that it complies with local practices and procedures. Many different things can vary, the type and size of wire required in a particular location, the number of circuits, outlets or lighting points – also you may be required to have your work inspected by either a licensed electrician or a building inspector before switching on.

Never begin any work you know, or think, may not comply with your local or National Electrical Code.

Always use U.L. listed materials, as anything else may create a fire hazard.

Always follow the manufacturer's instructions supplied with any materials, fixtures or appliances.

Simple improvements

The best way of working out how you can improve your existing system is to take a tour round your house. Look out for areas where there are a lot of electrical appliances and too few electrical outlets; extension cords trailing across the floor are often the tell-tale signs. Some appliances might benefit from being permanently hooked up to the power supply instead of just being plugged in; examples are freezers, waste disposal units and extractor fans, to name just three. Ask yourself whether outlets are conveniently located – some may be too low for easy access, or completely hidden behind furniture.

Look at the lighting too. If you have just a central fixture in each room, anything would be an improvement, and even if you are happy with what you have, replacing some of the fixtures could change and improve the lighting in the room concerned. Adding more lights, or altering the positions of existing ones, could be the answer. If it is ideas you need, keep your eyes open in public buildings – hotels, restaurants, banks and so on – here you will often be able to see quite advanced lighting ideas in use which can be more helpful than visiting a lighting showroom.

As you work your way round the house, note down what you have and an idea of what improvements are needed on a rough floor plan. This will be a great help in future when you come to plan and carry out the improvements you need.

WARNING

Electricity can be dangerous. Always check that the circuit is dead before beginning any work, either by unscrewing the fuse or by turning off the breaker and then double checking the circuit with a voltage tester. Always use quality materials with a U.L. listed label. After completing the job carefully re-check your work before switching the power back on. Remember also that, whenever you shut off your electricity supply, you should also check any instructions for your central heating system or your boiler before turning the supply back on.

Major alterations

If you have an out-of-date system – or your initial survey shows up really serious shortcomings including inadequate power supply, you will have to consider some major rewiring work. It is often easier to plan and install completely new circuits placing them exactly where you want them and which will provide all the facilities you need rather than spending hours tracing existing circuits and working out how to modify them. Of course, it will often be possible to use parts of the existing set-up if they are in good condition, but this is best looked upon as a bonus.

Apart from improvements to your existing lighting and power circuits, you may want to install extra circuits for major appliances such as waste disposal units, washing machines and clothes dryers, or to provide power to auxiliary buildings. This may well entail fitting a new service panel to cope with the extra circuits – a chance to ensure that this too is able to cope with all your needs.

How electricity works

Electricity reaches your home by an underground or overhead supply cable. This is tapped off the local street supply coming from a nearby substation. This is called "three phase" because it has three phase or live cables and one neutral. The voltage difference between any pair of live cables is 240 volts – the voltage normally supplied to heavy current users such as factories – while that between any of the phases and the neutral is 120 volts – the normal voltage for household use. Factories and other such industrial buildings draw current from all three phases, while the household supply can also be 2 phase, three conductors – two live or "hot" wires and one neutral.

With 120 volts available between either hot wire and the neutral and 240 volts between the two hot wires this provides the home owner with enough electricity to operate normal household circuits and also heavy appliances, such as ranges, which generally require 240 volts.

The service panel

When the supply cable enters your house, it passes first of all to the utility company's service head and from there to the electricity meter, which records how much electricity you use. Up to this point, everything belongs to the local utility company, and must not be tampered with in any way; the service head and meter are fitted with special seals so any interference can be readily detected.

From the meter, two more cables called meter tails run to your main service panel. In older homes there may be several fuseboxes, all linked to a main disconnect to which the meter tails are connected; in a modern house there will be a one-piece service panel which may also contain a main breaker switch as well as the individual circuit breakers.

Distribution

The service panel controls the distribution of current to the various circuits in the house. The live meter tails are connected to strips called busbars along which are mounted a number of individual fuses or circuit breakers of different amperage ratings, depending on the type of service panel. Their purpose is to protect each circuit from receiving too much current – either because the consumer tries to take too much, or if an electrical fault occurs – the fuse will "blow" (or the breaker will trip) and cut off the supply to that circuit. The fuse may be the replaceable screw-in type containing a length of fuse wire, or may contain a cartridge fuse with the wire concealed in a small metal-capped ceramic or cardboard tube. On the most modern installations the fuses may be replaced by small electro-mechanical switches that trip off if excess current is drawn by the circuit. In addition, the main supply circuit bringing power to the service panel is also protected by a main fuse or circuit breaker. Removing the fuse or shutting off the breaker turns off the power to the whole house.

The end of each fuse or circuit breaker is connected to the live wire of the circuit cable. The returning neutral wires of all the circuit cables are linked to a neutral terminal block to which the neutral meter tail is connected. Within the service panel you will also see that the ground wires from each circuit are also attached to the neutral busbar, and from there a single grounding wire runs from the neutral busbar to the grounding point in the house – it is attached to the incoming cold water supply pipe (metal and not plastic) or to a grounding rod driven into the earth beneath the house.

Individual circuits

From the fuses or breakers in the fusebox or service panel individual cables run out to supply the various circuits in the house. All circuit wiring is color coded – the neutral wire is always white, the grounding wire may be either green or bare without any covering, the "hot" or live wire in a 120 volt circuit is usually black or another color, but not white or green. To obtain 120 volts at an outlet requires connecting one black wire and one white wire. To obtain 240 volts

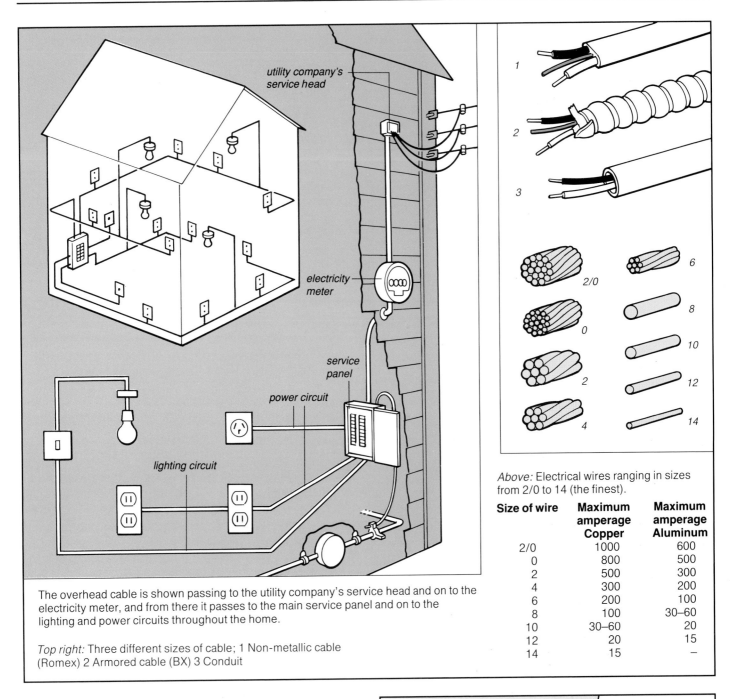

The overhead cable is shown passing to the utility company's service head and on to the electricity meter, and from there it passes to the main service panel and on to the lighting and power circuits throughout the home.

Top right: Three different sizes of cable; 1 Non-metallic cable (Romex) 2 Armored cable (BX) 3 Conduit

Above: Electrical wires ranging in sizes from 2/0 to 14 (the finest).

Size of wire	Maximum amperage Copper	Maximum amperage Aluminum
2/0	1000	600
0	800	500
2	500	300
4	300	200
6	200	100
8	100	30–60
10	30–60	20
12	20	15
14	15	–

requires connecting two black or colored wires – however the two black wires must be fed from opposite supply phases at your service panel. Circuits carry different loads depending on the job they are required to do, ranging from 15 or 20 amps for a normal domestic circuit up to 50 amps for, say, a stove or central air conditioner. The wires making up each circuit are sized according to the load that they carry; number 12 wire and number 10 wire are the most common in household work – (wire sizes get smaller as the gauge number increases) – number 14 gauge wire is the minimum size wire for most household work.

• CHECKPOINT •

MEASURING ELECTRICITY

Watts measure the amount of power needed by an appliance when in use, and is usually marked on its casing. One thousand watts (1000W) is equal to one kilowatt (1kW).

Amps (Amperes) measure the flow of the electrical power needed to produce the necessary wattage for an appliance.

Volts measure the "pressure" supplied by the generators of the local Utility Company that forces the current along the conductors to the various outlets.

Once you know the number of watts (or kilowatts) an appliance requires and its necessary voltage, you can find the outlet, plug and cable of the appropriate amperage to use.

CHAPTER 1
UNDERSTANDING THE BASICS

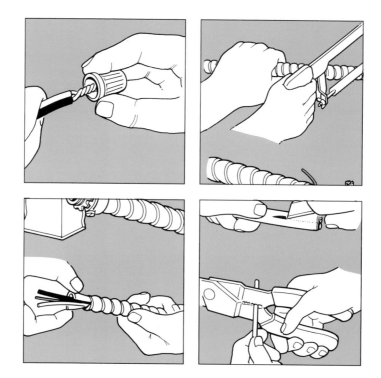

In order to achieve a thorough understanding of the way a home electrical system works, the following chapter explains exactly what happens to electricity after it has left the utility company and enters your house through the service panel and into the power and lighting circuits.

It illustrates the basic tools you will need to carry out repairs and installations, tells you how to use testing devices — emphasizing safety at all times. The project on preparing cable (page 15) explains the specific functions of the live, neutral and grounding wires, and how to join cable the safe and professional way. Information is also given on running cable with or without conduit, under wooden floors, over solid floors, chased into masonry walls or clipped to partition walls for the safest and neatest effects.

THE CIRCUITS

Normally both power and lighting are combined into common circuits, the only controlling factor being the maximum rated amperage load allowed for each circuit, protected by a similarly rated circuit breaker. The normal ratings of single circuits for domestic use are 15 or 20amps. For heavier loads, such as appliances or air conditioners which draw down a greater load, individual circuits will be dedicated to that particular use and are frequently rated higher (30amps). In special cases two circuits, each of 120 volts, may be combined to create a 220 volt circuit to supply greater power demand to ranges, for example.

Domestic lighting components
Domestic lighting circuits are usually of the radial kind and may be one of two following systems.

Junction box lighting In older homes, you may have a junction box system which incorporates a junction box for each light. These boxes are placed on the single supply cable at a convenient position between the ceiling outlet and the light switch on the wall. A cable runs from each junction box to the ceiling outlet and another from the junction box to the light switch.

Loop-in lighting The most widely used lighting circuit is the loop-in system. This has a single cable that runs from ceiling junction box to ceiling junction box and finishes at the last one on the circuit, while an additional single cable runs from each ceiling junction to the light switch. Loop-in circuits are arranged one to each floor, and are protected by separate fused circuits in the main service panel.

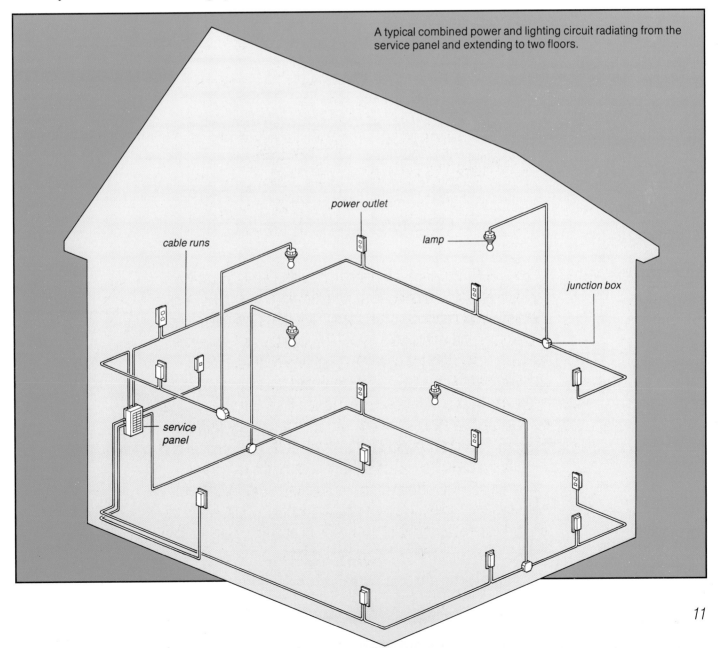

A typical combined power and lighting circuit radiating from the service panel and extending to two floors.

power outlet

cable runs

lamp

junction box

service panel

Power

The most popular form of power circuit for feeding outlets is the ring circuit in which a cable starts from the terminals in the service panel and is taken around the house connecting one outlet to another until it arrives back at the terminal, to complete the ring. By this technique, power reaches all the outlets from both directions which reduces the load on the main cable. Like lighting ring circuits, power ring circuits are usually arranged one to each floor.

The number of outlets on a ring circuit can be increased by adding extensions called "spurs". A spur can be a single cable connected either to the cables of an existing outlet or from a junction box inserted within the ring.

Taking power to auxiliary buildings

The power supply to give lighting and power outlets to a separate garage or workshop, for example, will require its own circuit.

The cable must run from its own fused circuit in the main service panel and pass safely underground or overhead to the outside building, where it is wired into another switch fuse subpanel from which the new lighting and power circuits can then be distributed as required (see page 57).

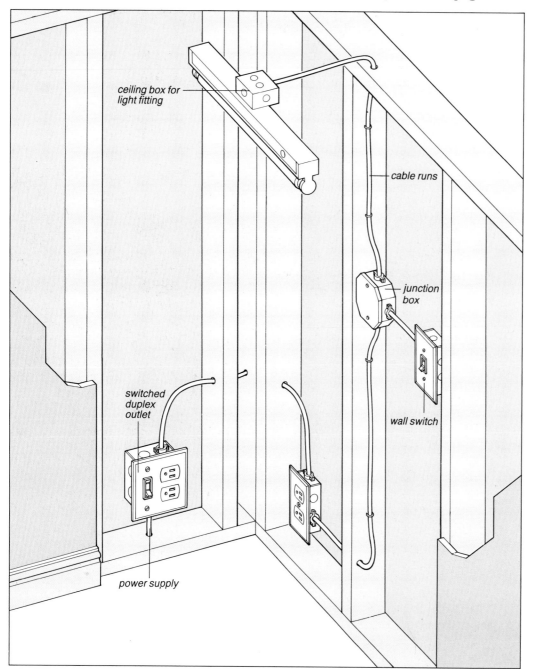

A typical lighting and power circuit extension showing a duplex outlet (15/20A) at baseboard level with a light fitting and wall switch above.

ceiling box for light fitting

cable runs

junction box

switched duplex outlet

wall switch

power supply

TOOLS AND EQUIPMENT

Although very few specialist tools are needed for making electrical connections, several general purpose tools will be required for lifting floorboards, drilling and cutting into ceilings and walls.

Terminal screwdrivers (both large and small). Choose those that have plastic handles and plastic insulating sleeves on their shafts.
Wire cutters (Electrician's pliers). These are fitted with an insulated sleeve on the handles.
Diagonal cutters, also similarly insulated, will cut through thicker cables and wires than wire cutters.
Small hacksaw This may be needed for cutting through the thickest of cable wire.
Wire strippers have insulated handles, and jaws specially shaped to cut through the plastic insulation sheathing of cable without damaging the wire core.
Cutting knife Choose a knife with razor-sharp disposable blades for cutting and removing the outer covering of cable.

Power drill A power drill is recommended for making holes through timber, brick and plaster for which you will need the correct types and sizes of bit.
Fish tape This is useful whenever you need to pull wires or cables behind walls or through conduit. Made from a length of flattened spring steel wire, 25/50 feet long, and comes on a reel for easy handling.

The following general purpose tools are also recommended.
Needle-nose pliers
Claw hammer (with an insulated handle),
Club hammer (for use with a cold chisel and bolster)
Cold chisel
Bolster chisel
Wood chisels
Floorboard saw or tenon saw
Spirit level
Steel measuring tape
Bradawl
Torch

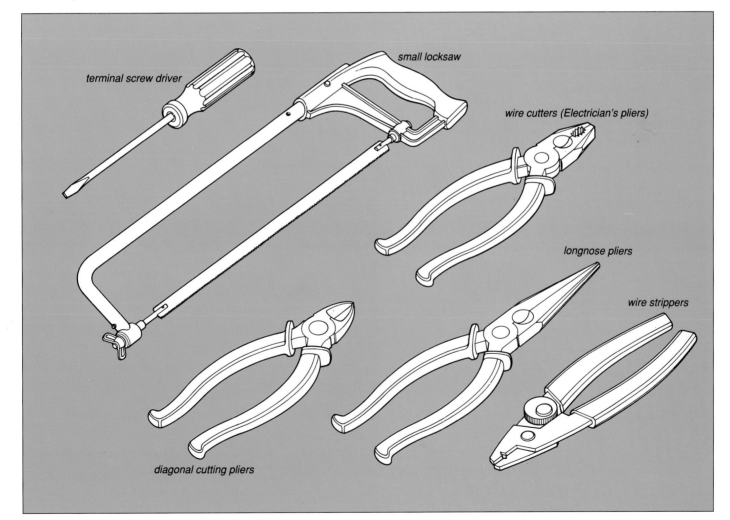

terminal screw driver

small locksaw

wire cutters (Electrician's pliers)

longnose pliers

wire strippers

diagonal cutting pliers

TESTING DEVICES

There are two basic diagnostic tools available which are essential in all electrical work.

The neon voltage tester

This instrument can be used to determine whether or not the circuit is hot. To carry out this operation begin by holding the tester leads only by the insulated areas, touch one probe to a hot wire or terminal and the other to the neutral wire or terminal. If the tester lights up, the circuit is hot.

You can also use this device to determine which is the hot wire of two-wire circuit with ground. In this case, touch one probe to the grounding wire or metal box and then touch the other probe to the other wires in turn. The tester will light up when it touches the hot wire.

When using the tester, be very careful at all times. Do not touch any metal parts with your hands. And remember that a carelessly placed probe can cause a short circuit if it accidentally touches both a hot and a grounded object at the same time. These instruments are small and can easily be covered up accidently.

Continuity tester

Continuity testers are available in several forms: one contains a battery and a light and the other uses a battery and a buzzer. Either type is used to test whether a circuit is open or broken, or whether a short circuit exists. Before using either device, make sure the power is off by turning off the breaker or pulling out the fuse.

The following example explains how to use the tester. After completing the circuit from the distribution center to a new light fixture, you wish to check the current before turning on the power. Attach the alligator clip of the tester to the neutral busbar and touch the end of the hot wire with the tester probe. Without a light bulb in the new fixture, get someone to turn the switch on and off. If the tester does not light or buzz, you know that the hot wire has not shorted. Now put the light bulb in the fixture and get your assistant to repeat the switching. The tester should light or buzz with the switch on. Now you can put the new hot wire into the circuit breaker terminal and energize the circuit.

The continuity tester can also be used to determine whether or not a cartridge fuse has blown.

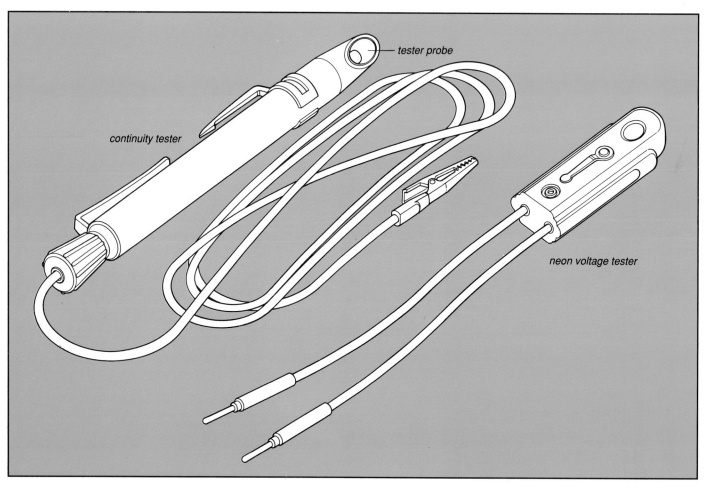

tester probe

continuity tester

neon voltage tester

PREPARING CABLE

To join Romex cable, split the plastic covering about 8in from the end with a sharp blade, being careful not to cut any of the conductors. Remove any paper wrapping from around the conductors, strip about ¾in of the insulation on each wire with a pair of electrician's wire strippers. Insert the end into a connector made for plastic cable and tighten the clamping screws. Push the connector into the junction box and tighten the lock nut.

To join BX cable, use a hacksaw to cut about 8in from the metal armor. Hold the cable securely and cut the armor across – not with – the spiral. When the cut is nearly through hold the armor above and below the cut and twist sharply. Insert a plastic bushing around the wires into the end of the cable to protect the wires from the sharp edges of the armor. Pull the grounding wire away from the others and wrap it around the armor (in this system the armored covering to the cable acts as a ground). Attach a BX connector onto the end of the cable, push it back firmly and tighten the screw. Push the connector into the box and tighten the locknut.

Junction boxes

All connections between the circuit wiring and any switch, outlet or fixture, or any splices in the circuit wiring must be made within a junction box. Under no circumstances must electrical wire be spliced with tape and left exposed. Junction boxes come in three shapes – octagonal, square, and outlet boxes. The three shapes of box come in a variety of depths and a variety of types depending on the proposed location. The outlet box is used for switches and receptacles, which can be ganged by removing one side and joining it to another to make a double box, or more.

Outlet boxes are mostly attached directly to wall studs by means of a flange, or set into paneling or drywall between the studs by use of clamps or clips. The hexagonal box is most often used for hanging electrical fixtures and is attached directly to the studs or ceiling joists. The square box is most often used to splice wire together. Individual wires to be spliced are held together with a wire nut, which securely locks and twists the wires together. Note that white wires are neutral, green or bare copper wires are ground. Black or red (or any other color except white or green) are live wires.

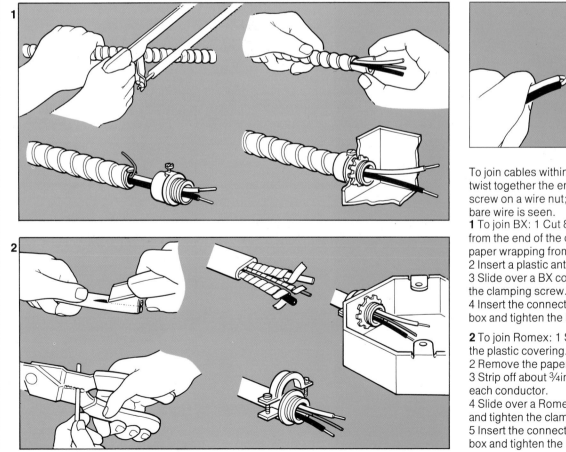

To join cables within a junction box, twist together the ends of the two wires, screw on a wire nut; tighten until no bare wire is seen.
1 To join BX: 1 Cut 8in of the armor from the end of the cable and remove the paper wrapping from around the wires.
2 Insert a plastic anti-short bushing.
3 Slide over a BX connector and tighten the clamping screw.
4 Insert the connector into the junction box and tighten the locknut.

2 To join Romex: 1 Strip about 8in from the plastic covering.
2 Remove the paper wrapping.
3 Strip off about ¾in of the insulation on each conductor.
4 Slide over a Romex cable connector and tighten the clamp.
5 Insert the connector into a junction box and tighten the locknut.

•CHECKPOINT•

RUNNING CABLE

Two kinds of wire are available: non-metallic sheath cable (Romex) which is a flexible cable covered in plastic, or metal armored cable (BX) which has a flexible steel spiral covering. For new projects always try to use the same type of wire and equipment as already exists in your home, as this has most likely been determined by your local electrical code. In addition to these two wiring systems, individual insulated conductors can be run in galvanized steel pipe or conduit, which may be rigid (threaded) or thin wall (unthreaded). Conduit can be obtained in a variety of sizes and comes in 10 foot lengths – and must be bent by using a conduit bender. The code requires that conduit be used in certain situations.

With wooden floors, you can run circuit cables underneath them – or in the void between the floor and the ceiling above.

Under first floors, you can let the cable rest on the concrete slab, although it is better to clip it to the sides or bottom edges of the joists if possible. At second-floor level you can let the cable rest on the ceiling surface if it runs parallel with the joists; if it crosses the joists line, drill small holes in the joists – ideally at their mid-depth point – and thread the cable through the holes. In this case, lift only one floorboard and run the cable right across the room; with cable runs parallel to the joists lift a board at each side of the room and "fish" the cable through.

Solid floors

If you have solid floors, you can cut out a channel and run in the cable, which should be protected by conduit, but it is generally a lot simpler to run the cable in plastic raceway mounted on top of the baseboard or even behind the baseboard.

In attics you should clip the cable to the joists above the level of any insulation. Thread it through the joists where there are walkways.

Cables in walls

If you have masonry walls, you can bury the cables in shallow channels (called chases) cut in the wall surface with a mason's chisel and light sledge hammer or a rented chasing machine. You can lay the cable in without any further protection, but it is better to either thread it through protective conduit or cover it with pin-on protective channeling, even though both methods mean you need a wider chase. The chase is simply plastered over when the cable is in place.

With stud partition walls, you can thread the cable down between the sheets of drywall on either face, but you will have to drill through the head- and sole-plates and also expose any bracing so you can cut a small section away and allow the cable through.

For surface-mounted cable, either clip it to the wall, the baseboard, door and window frames by running it in a metal or plastic raceway. This has a base which you pin, screw or stick to the wall and a snap-on cover to conceal the cables.

Running underfloor cables parallel to the joists; a flat-section sprung-steel coil helps to feed the cable through.

Running cables across the joists; drill clearance holes at least 2in below the top and central to the floorboard.

Plastering over plastic conduit, chased into the wall, through which the cable has been threaded.

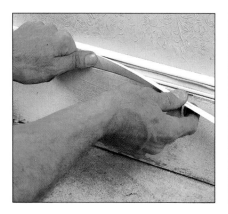

Running cable through mini-trunking; nail or stick the channel to the wall, feed in the cable and snap on the cover.

Nailing a covering channel over cable buried in a chase in the wall to provide some protection against stray nails.

Fixing cable to a door frame with plastic cable clips; these are available in sizes to match the various cable dimensions.

CHAPTER 2
SIMPLE REPAIRS

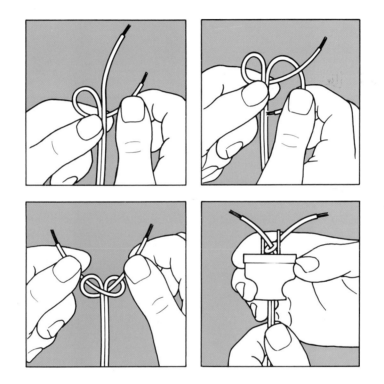

It not only makes economical sense to be able to do simple electrical repairs and maintenance around the home (see page 16), but if they are carried out when they are needed your home will be that much safer for everyone who uses it.

This chapter explains how to deal with blown fuses and tripped breakers, how to make simple repairs to lampholders and replace damaged switches, how to fit three- and four-way switches — a relatively simple operation which means that the lighting in places like staircases can be operated from switches on all floors. Information is also given for fitting dimmer switches so that you can control the lighting ambience in any room you wish.

REPLACING SIMPLE FITTINGS

Many light fittings and appliances are supplied with electricity through their attached flexible cords that are simply plugged into the system and are as easily disconnected. Always remember to remove the plug from the outlet before working on the equipment. Do not just switch off the appliance as this does not ensure that the electrical current is isolated.

Plugs

There are several different shapes of plug – some are round and others are oblong and specially shaped to fit into outlets and adaptors. While certain plugs need to have their wires stripped and fixed to terminal screws, where the underwriter's knot is used (see below), others are attached to the cord simply by clamping within the casing of the plug.

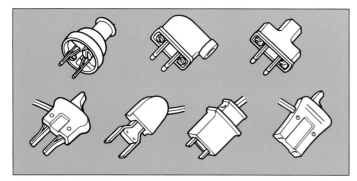

Typical plugs showing the range of size and variety.

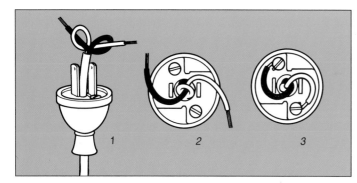

Certain plugs attach to the cord without first having to strip the wires. 1 Lift clamp, slide in cord, close clamp. 2 Hold the prongs and remove case, slide cord through case. Open prongs, insert wire, close prongs and slide case back on. 3 Remove case, insert wire and push case back on.

The underwriter's knot

This is the knot electricians use in a plug or lampholder to prevent the flexible cord from becoming strained when the attachment is pulled in everyday use – where, obviously, any loose wiring within a plug would be potentially dangerous.

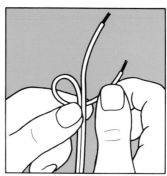

To tie an underwriter's knot:
1 Unzip the cord about 2in.

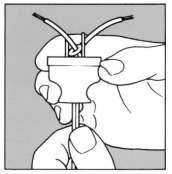

2 Make the first loop, passing the end behind the cord.

3 Make a second loop around first free end. Pass second free end through first loop.

4 Pull the free ends tight. Slide the cord through the plug or lamp holder.

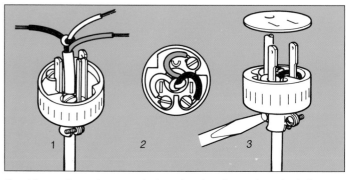

Rewiring a plug using round wire, showing the underwriter's knot.

Grounded 3-prong plug. 1 Slide wire through plug and tie under-writers' knot with black and white wires. 2 Pull wire tightly into plug, twist free ends clockwise, and attach them to terminal screws. 3 Tighten clamp on cord and replace fiber cover over prongs.

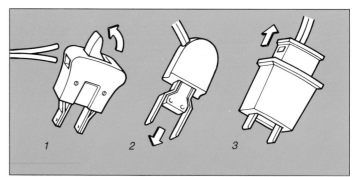

Round wire plugs. 1 Slide wire through plug and tie underwriter's knot. 2 Pull wire tightly into plug and twist free ends clockwise. 3 Attach ends of the wire to terminal screws.

FUSES AND CIRCUIT BREAKERS

The purpose of a fuse or circuit breaker is to protect you and your home in case of a fault in the electrical system. A fault may be due either to a "short circuit" (a live wire coming into contact with a ground or neutral wire) or due to overloading of a circuit. These actions open the circuits, thus disconnecting the supply of electricity. It is necessary to determine the cause of any fault before replacing a blown fuse, or re-setting the circuit breaker at the service entrance panel (or in some cases at a sub-panel).

If a fuse blows when a particular electrical appliance is switched on then it is probable that there is a short circuit within that appliance. Alternatively, there may be several other appliances already on that circuit and the power needed for the additional appliance may have caused an overload. A black or discolored window on the fuse indicates a short circuit; if however the metal strip is broken it is likely that the circuit was overloaded. In the case of a tripped breaker a red indicator square is automatically uncovered to help identify the faulty circuit.

WARNING

Before attempting to carry out any electrical repair to permanently wired equipment, such as switches and power outlets, turn off the electricity supply at the mains switch on the service panel and remove the fuseway to isolate the appropriate circuit. Then, using a voltage tester, double check to make sure the terminals are dead.

Types of fuses

It is advisable to keep a supply of fuses of the correct rating and type close to the fuse panel. Never replace a blown fuse with another of a different rating; fuses of 15 amp, 20 amp and 30 amp are the most common in domestic circuits. For adequate protection, the amperage rating of a fuse or breaker must be the same as that of the circuit conductor it protects. For example, a circuit using number 12 copper conductor has an ampacity of 20 amps (see page 9); the fuse or circuit breaker therefore, must also be rated 20 amps.

The regular plug fuse has a base similar to that of a light bulb and screws into the fuseway in the fuse panel. A metal strip shows through the clear mica panel on the front of the fuse and indicates whether the fuse is good or not.

The time delay fuse is almost identical to the regular fuse except that it allows temporary overloading on the circuit which is useful to allow for the power surges in such appliances as washing machines and dish washers.

The Type "S" fuse prevents use of a wrong size fuse by means of an insert which screws into the fuseway accepting only a fuse of the correct rating.

To replace a screw-in type fuse, first shut off the power by withdrawing the main fuses, identify the fault on the circuit and make the necessary repairs, unscrew the blown fuse and screw in a new one of the correct type and rating, finally restore the current.

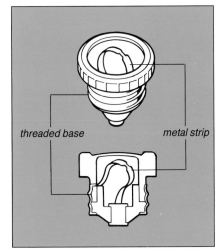

The standard plug fuse has a base similar to that of a light bulb and screws into the fuseway. The metal strip is visible through a window on the front of the fuse.

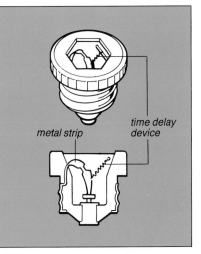

The time delay fuse allows for temporary circuit overloading, useful on circuits containing electrical appliances which draw a heavy initial current.

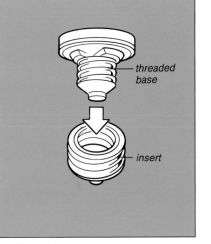

The Type S fuse allows only the fuse of the correct rating to be inserted into the fuse holder.

SIMPLE REPAIRS

Some fuse panels may be equipped with cartridge type fuses. It may be difficult to detect which of these fuses may have blown. To replace a cartridge type fuse first shut off the main power, remove the fuse by means of a fuse puller and check the fuse with a continuity tester to see if it has blown. If it is faulty replace with a new fuse of the correct type and rating.

Circuit breakers do the same job as a fuse by means of a small electro-mechanical switch which trips to the off position in the event of a short circuit or an overload. A red rectangle is exposed on the breaker when it is in the tripped position which helps to identify the faulty circuit. Unlike fuses, which work on a self-destruct principle, circuit breakers can be reset (turned back on) once they have tripped. To reset a breaker, first identify and correct the fault; push the breaker fully to the off position which will reset the switch, then push fully to the on position.

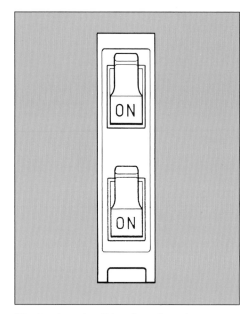

The tandem circuit breaker allows two circuits to be taken off in the space of a standard size breaker.

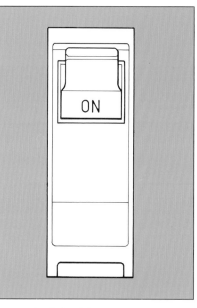

The standard circuit breaker. To reset a tripped breaker push the switch fully to the off position before pushing to the on position.

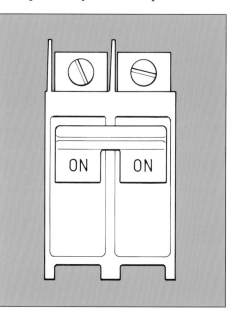

The double pole breaker is for use with 240 volt appliances and ensures that both phases of the supply to the appliance are tripped in the event of a fault on one phase.

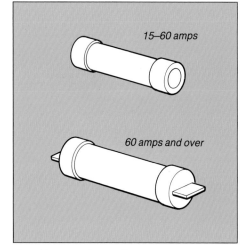

Cartridge type fuses. The round types are rated from 15 to 60 amps. The knife edge types are rated from 60 amps and over.

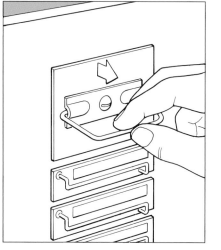

To remove a main fuse, pull out the plastic fuse carrier from the fuse panel.

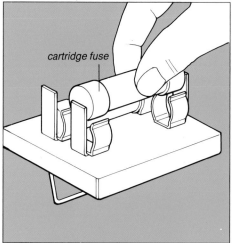

To replace a main fuse, pull the fuse carrier from the fuse panel, remove the fuse from the carrier and replace with a new fuse of the correct rating.

1

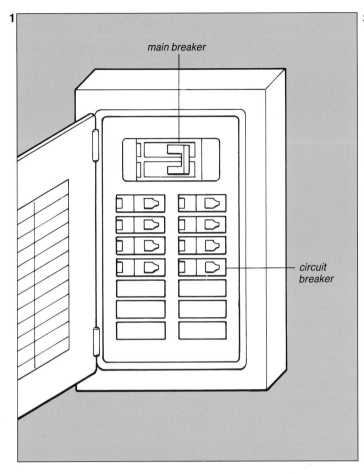

main breaker

circuit breaker

3

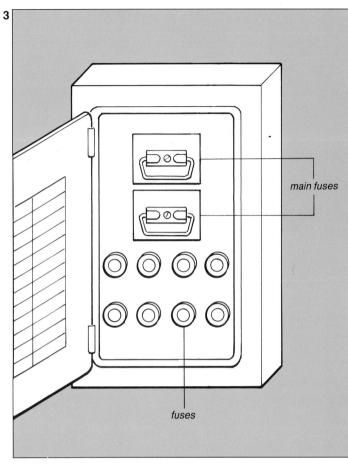

main fuses

fuses

2

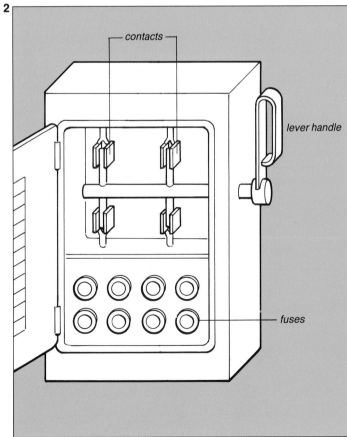

contacts

lever handle

fuses

1 Circuit breaker panel. To shut off the power turn the level handle to the off position.

2 Lever type disconnect. To shut off the power turn the lever handle to the off position.

3 Fuse panel. To shut off the power pull out both of the main fuse carriers.

REPLACING A CEILING CANOPY

Switch off the power at the appropriate circuit breaker to isolate the circuit. Remove the existing light and fitting and the canopy and, using a voltage tester, double check to make sure that the terminals are quite dead.

After removing the canopy, it is important to identify the wires and, perhaps, make a careful diagram of the way the wires are connected, so that you can confidently rewire the new ceiling junction box in the same way.

Fixing the new canopy

Disconnect the wires from the terminal canopy. If the existing junction box is unstable you might want to nail a mounting board between the joists above the box position and drill a hole through for the cable. Screw the backplate in position and bring the cable through.

Make sure the ends of the wires are sound and clean, then fix them to the terminals carefully following your diagram. Having made sure the terminals are soundly connected, thread the new canopy up the cable and screw into the existing junction box.

REPLACING PENDANT LAMPHOLDERS

Lampholders and their cables are subject to a considerable amount of heat from the light bulbs and should be checked periodically for damage. The most common types of holder are metal and plastic and are designed to take only two-core cable.

In time, plastic holders may become brittle and more easily cracked or broken. The metal contacts inside the holder may also become corroded which would prevent them from making a good electrical contact.

Begin by unplugging the lamp, or if permanently wired, either remove the fuse or trip the breaker off to isolate the circuit, as described in the warning box opposite. Unscrew the cap of the lampholder and slide it up the cable to expose the terminals. Loosen the screws and pull out the wires. Should the wires be broken, cut back slightly and expose core before fitting the new holder. Slide the cap of the new holder well up the cable and hold with electrician's tape. Expose sufficient length of wire and make the underwriter's knot (see page 18), to take the weight off the terminals, and screw on the cap.

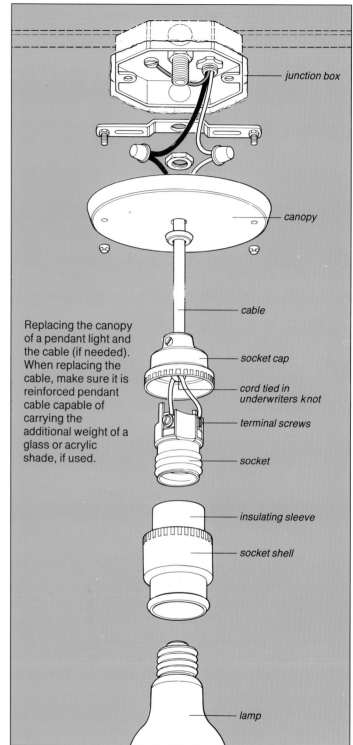

junction box

canopy

cable

Replacing the canopy of a pendant light and the cable (if needed). When replacing the cable, make sure it is reinforced pendant cable capable of carrying the additional weight of a glass or acrylic shade, if used.

socket cap

cord tied in underwriters knot

terminal screws

socket

insulating sleeve

socket shell

lamp

To fit the new holder, thread the cable through the socket cap and remove 2½in insulation. Tie the underwriter's knot. Tightly twist each stranded wire clockwise, wrap one wire clockwise around each terminal screw and tighten the screws.

WARNING
Always match fuse size to circuit rating. For instance, if the circuit is rated 15 amps, use a 15 amp fuse, not 20 amp.

REPLACING A LIGHT SWITCH

Replacing a damaged switch is a relatively simple operation. All that is required is that the existing wiring is connected to the new switch in exactly the same way as it was connected to the old one.

Turn off the power and remove the fuseway to isolate the circuit before removing the faceplate and switch. Then use a voltage tester to double check that all the wires to the switch are completely dead. Make sure the new faceplate and switch fit the old switch box, otherwise you will have to replace all parts (see page 24).

A one-way switch will be serviced by two conductors and a ground cable, which should be connected to a ground terminal on the mounting box. The black and white wires should be connected to the switch itself.

The back of the new faceplate will be marked to indicate the top so that when you replace it the rocker will be depressed when the light is on. Where at all possible, use the old screws and in this way, you will know that the threads match and will ensure a neat, secure fit.

Replacing a one-way light switch and faceplate.

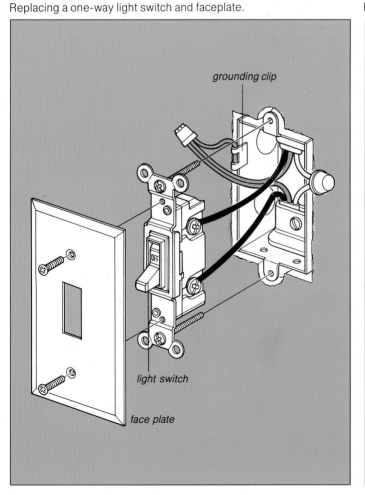

REPLACING A RECEPTACLE

To replace a damaged receptacle with a similar type is quite a straightforward operation. Begin by turning off the power supply and removing the fuseway to isolate the circuit, then, using a voltage tester, double check the wires to be sure they are dead.

Remove the fixing screws holding the receptacle and ease it out of the box. Make a careful note, or diagram, as to how each wire is connected and then loosen the terminals and free the wires. Check that the incoming wires inside the box are perfectly sound, and then connect the wires to the terminals of the new receptacle in exactly the same way as the old wires, following your own diagram or notes. Tighten any unused screw terminals. Note that this type of back wiring is suitable only for copper wire.

Carefully push the wires and receptacle into the box. Screw the device to the box and then screw on the faceplate to complete the replacement. Turn on the power supply.

Replacing a damaged receptacle and faceplate.

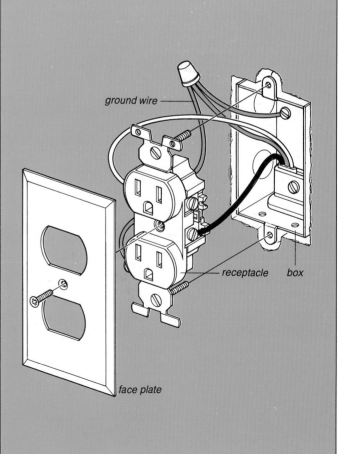

FITTING SWITCHES

Most lights – ceiling or wall fixtures – are controlled by a separate switch mounted on the wall at a convenient position. The usual site is next to the door just inside the room at about 4ft above the floor. The switch cable from the light (or the junction box providing the switching connection) runs down to the switch position (usually taking the shortest route), and is connected to the switch terminals.

Wall switches can be surface- or flush-mounted and need a mounting box depending on the type. Metal flush boxes are screwed into a recess cut in the wall;

A pair of end-of-the-run wired switches.

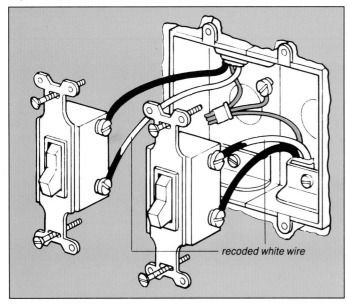

recoded white wire

A switch wired in the middle of the run.

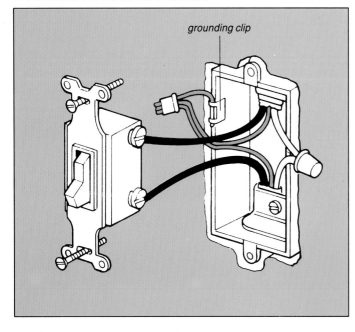

grounding clip

the plastic surface-mounting ones are screwed to the wall surface, and are often used in conjunction with surface wiring in plastic or metal raceways.

The most common type is the one-way switch called a toggle switch or a single pole switch; each switch usually has two terminals labeled L1 and L2, to which the wires of the switch cable are connected (the cable ground core usually goes to a ground terminal on the switch).

Three-way switches have three terminals each, two traveler terminals and one common terminal. These switches operate in pairs to provide control of one light from two different switch positions – known as three-way switching – in this case, linked by special three-core and ground cable. The three cores are color-coded red, black and white for identification. Switching from three (or more) positions is possible using one or more intermediate switches between two three-way switches; these have two pairs of terminals; known as four-way switches; (see page 27).

WARNING

Before commencing any work:

● Make sure that you are completely familiar with the relevant requirements of your local Electrical Code. Note that electrical codes vary from one location to the next.

● Always make sure that the power is turned off.

● If ever you are in any doubt consult a licensed electrician or your local Building Inspector.

● Check your home owner's insurance policy – some are invalidated if any electrical work is carried out by anyone other than a licensed electrician.

When carrying out any work:

● Double check that a circuit is off by using a voltage tester.

● Use only U.L. listed and approved parts and materials and install them in accordance with the requirements of your local Electrical Code.

● Use proper insulated tools and as a precaution stand on a rubber mat. Use a wooden ladder not a metal one.

● Always double check work before restoring the current.

● Have your work checked by your local Building Inspector or a licensed electrician if required by your local Code.

THREE-WAY SWITCHES

There are two areas of the house where three-way switches is particularly useful – in the bedroom, where you may want to turn out bedside lights without having to get out of bed, and on the stairs, so you can control both hall and landing lights from top or bottom of the stairs. The hall/landing set-up is a classic example of three-way switching at its best.

For full control of both lights from both locations, you need a pair of three-way switches on the landing and another pair in the hall. The left-hand of the two switches is linked by one three-wire and ground cable; another similar cable links the two right-hand switches. The final wiring depends on the location of the fixtures and of the power source. There are several possible situations. The power supply can be at the beginning of a run while the fixture is either at the beginning, middle or end of a run or the power and the fixture can both be in the middle of the run (see diagrams below).

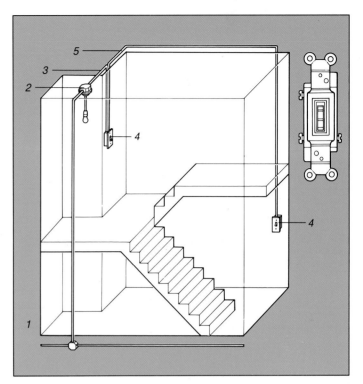

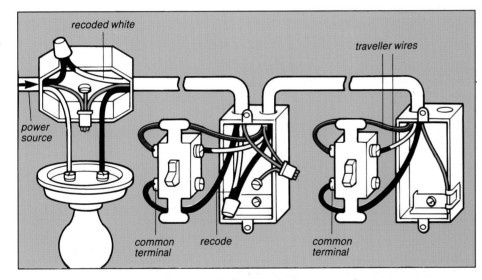

Three-way lighting circuit
1 Power source from circuit
2 Light fitting with ceiling junction box
3 Two-wire with ground
4 Three-way switch with junction box
5 Three-way wire with ground

The three-way switch is identified by three terminals and a plain toggle. These switches operate in pairs to control light or receptacle from two loctions.

Power source and fixture at the beginning of the run.

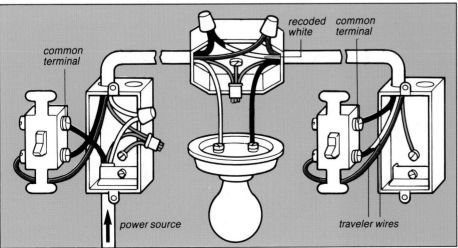

Fixture at the middle of the run.

Another advantage of this arrangement is that if the landing and hall lights are on different circuits you will always have light on the stairs (vital to prevent falls) even if one lighting circuit fuse blows. You must remember, though, that if one circuit blows, the other may still be live, so you must always isolate both circuits by removing the circuit breaker at the main service panel if you want to work on either the switches or any of the wiring.

If you have partial three-way switching (it is quite common to find installations where the landing light has three-way switching but the hall light does not) it is quite a straightforward job to upgrade it to full three-way control.

In bedrooms, several three-way switching options are possible. The simplest arrangement is to have switches to control the main light both by the room door and by the bed. The switch drop from the light goes to one switch or the other (usually whichever is nearer to it) and the two switches are linked as usual with three-wire and ground cable.

A more common situation is where a main ceiling light is switched by the door (via a single pole switch) while two individual bedside lights are controlled by switches at the door and at each side of the bed. This offers the maximum flexibility, since each bedside light can be switched on or off from the door or bed quite independently – an advantage when one partner is a late reader!

Such a set-up is obviously quite involved to wire up, and could, in principle, use a great deal of cable (depending on the dimensions of the particular run), especially the more expensive three-wire and ground variety. However, for most people, the positive advantages would probably outweigh the cost.

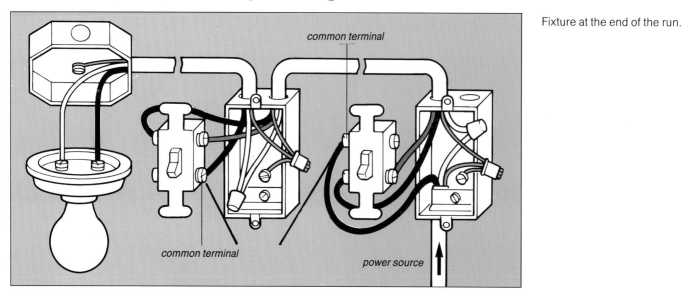

Fixture at the end of the run.

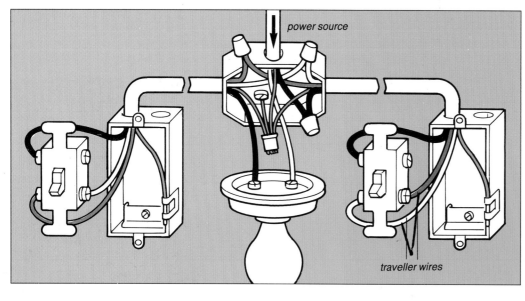

Power source and fixture at the middle of the run.

> **WARNING**
>
> Always turn off the power at the service panel before beginning any work. Double check that the circuit is dead by means of a voltage tester.

FOUR-WAY SWITCHES

You can control a light from three places by adding an intermediate switch to the circuit as described for a three-way switch (see page 25). The third switch interrupts the three-core and ground cable connecting the other two switches. It has two L1 and two L2 terminals.

At the mounting box you will have two identical sets of wires – red, black, white and green. Connect the green wires to the ground terminal on the box, and join the two red wires (which play no part in the intermediate switching) with a plastic block-connector. Ease the block to one side to clear the switch when it is fitted.

Connect the black and white wires of each cable to the L1 terminals on the new switch and those of the other cable to the L2 terminals. Screw the switch-plate to the mounting box.

Four-way lighting circuit
1 Power from source circuit
2 Light fitting with ceiling junction box
3 Three-way cable with ground
4 Four wire cable with ground
5 Three-way switch with junction box
6 Four-way switch with junction box

The four-way switch is identified by four terminals and a plain toggle. This type of switch is used only in combination with a pair of three-way switches to control light or receptacle from more than two locations.

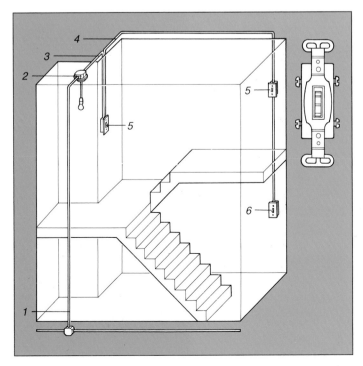

FITTING A DIMMER SWITCH

Dimmer switches are light switches that enable you to raise or lower the intensity of the light they control as well as providing the usual on/off switching. There are several styles: some have separate on/off switches and rotary dimming controls; some combine the rotary dimming control with a push on/push-off action; the most sophisticated are controlled just by finger pressure on the faceplate, or even by remote control. Most models are one-gang versions, often capable of three-way switching, but some include models with up to four gangs. Many include a small fuse to protect the delicate dimming circuitry from the effects of current surges.

You can fit a dimmer in place of an existing switch by removing the faceplate, disconnecting the switch cables and reconnecting them to the dimmer faceplate following the manufacturer's instructions.

There are two points to note. Firstly, many dimmers have a minimum and a maximum operating wattage and some models may not control a single bulb very well. Secondly, some need a deeper-than-usual mounting box, so check before you buy.

How to connect a dimmer switch, with middle-of-the-run wiring shown.

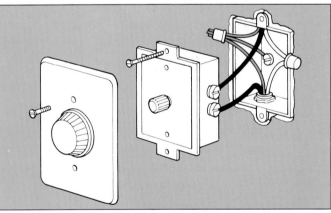

Connecting a dimmer switch in place of a switchplate; if it is for a fluorescent tube, check first that the switch is suitable.

CHAPTER 3
ADDING EXTRA LIGHTING

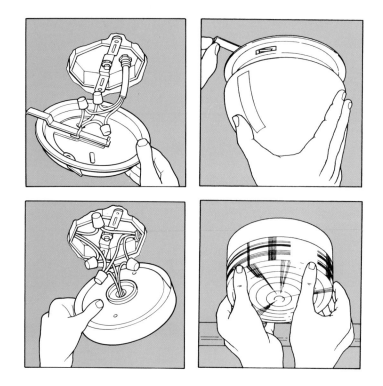

You may wish to add lighting around your home using the existing circuits. Depending on what the particular situation demands, you may decide to install a retractable fitting over your dining table, fit a fluorescent light in the kitchen, or create a more dramatic effect in your living room with concealed lighting or wallwashers. The choice of lighting fixtures is wide and varied ranging from simple pendant lights to track, bowl and close-ceiling fixtures.

The following pages show you how to install such lights, and to plan your home lighting schemes so that you will be able to create both practical and decorative effects to suit all situations and tastes.

EXTENDING LIGHTING CIRCUITS

If you want more lighting points around the house, or you want to move existing lights to new positions, you can do so simply by extending one of the existing lighting circuits (see page 11).

All circuits are wired up as radial circuits, with one circuit cable leaving the fuseway or the circuit breaker in the service panel and running out, feeding points along the way, until it reaches the most remote point on the circuit. With a 15 amp fuse and a supply voltage of 120 volts, such a circuit can, in theory, supply $15 \times 120 = 1800$ watts of power; each lighting point is rated nominally at between 60–150 watt, so up to 12 points can be supplied. In practice this is usually reduced to eight, to avoid overloading if light fixtures with more than one bulb are used. So the first step you have to take is to find out how much load you already have on each circuit.

To do this, turn off the main switch at your service panel and remove one of the 15 or 20 amp circuit fuses (or trip off the circuit breaker). Then go around the house and check how many lights and electrical outlets fail to work on that circuit. The next step is to determine the total wattage of all appliances and fixtures normally used on that circuit. This can be done by checking the information plate on each appliance or light fixture. The reserve capacity can then be found by deducting the present circuit wattage from the maximum circuit capacity (1800W for a 15A circuit or 2400W for a 20A circuit). If by adding an extra fixture the circuit wattage comes up to within 20% of the maximum circuit capacity – it would be better to check for other circuits with a larger reserve.

What type of circuit?

Assuming that you find you can extend an existing circuit, your next job is to work out what type of wiring practice has been used to wire it up, so you can decide how to carry out the extension work.

The new light position

You now have to decide on where the new light is to be positioned, so you can plan the best point at which to connect its power supply into the existing circuitry. If you are installing a new ceiling light downstairs, you may have to lift one or more floorboards in the room above so you can see where the circuit cables go and to enable you to run in the cable to the new light; the exception to this is if your new light is between the same pair of joists as an existing one, in which case you may be able to feed in the new cable between the existing and new light positions simply by fishing it along between the two holes in the ceiling. Fitting new ceiling lights upstairs is much easier so long as you have an attic above, since the existing cables can be run across the attic floor.

Once the position for the new light and switch has been determined, any convenient outlet will do as a source for the power to the new switch and fixture.

There are several possible power sources; a middle-of-the-run outlet, an end-of-the run outlet, a middle-of-the-run switch or fixture for you to choose from (see pages 45, 46).

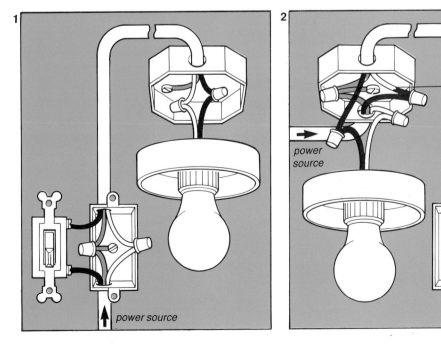

recoded white wire

power source

power source

1 Wiring for end-of-the-run fixture, middle-of-the-run switch.

2 Wiring for middle-of-the-run fixture, end-of-the-run switch.

WIRING A NEW LIGHT

Once a source of power has been identified, merely run a 2 conductor cable with ground to the nearest of the two new boxes – either the switch box or the fixture box, then run cable between the two new boxes. The following two situations can occur:

1 The power feeds directly into the fixture box. In this case the fixture is said to be wired in the "middle-of-the-run" and the switch is wired at the "end-of-the-run".
2 The power feeds into the switchbox. In this case the switch is said to be wired in the "middle-of-the-run" and the fixture wired at the "end-of-the-run".

Middle of the run fixture, end of the run switch
Three connections are necessary here.

1 The incoming neutral white wire is connected to the white wire from the new fixture.
2 The incoming "hot" wire is connected to the outgoing white wire on the switch loop; so this white wire must be recoded by wrapping it with electrical tape to show that it is now a live wire.
3 The returning black wire on the switch loop is connected to the black wire on the new fixture.

Cut a short length of grounding wire and splice together with all the other ground wires; attach the loose end to the fixture by means of a metal screw. At the switch box, recode the incoming white wire to show that it is a "hot" wire. Connect both wires to the switch terminals. Attach the ground wire to the switch box by means of a metal screw.

End of the run fixture, middle of the run switch
Two connections are necessary at the fixture, and three at the switch.

At the fixture box splice black to black and white to white and, again, cut a short length of ground wire and splice together with the other ground wires. Attach the loose end to the fixture by means of a metal screw.

At the switch box three connections are necessary:

1 Splice together the two neutral white wires.
2 Connect the two black wires to the switch terminals.

Cut a short piece of ground wire and splice it together with the other ground wires, then attach the loose end to the switch box with a metal screw or clip.

WARNING
Always turn off the power at the service panel before beginning any work. Double check that the circuit is dead by means of a voltage tester.

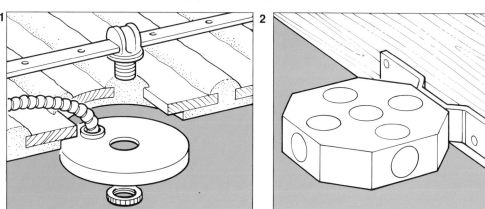

1 For existing ceilings, mark around the box, carefully cut a hole and fish through the wires. Insert a hanger bar and then attach the box to the bar.

2 Box with a flange for attaching to the side of a joist.

3 For positioning the fixture between two joists, use a box with an adjustable hanger bar.

FIXINGS TO CEILINGS

Whatever type of light fixture you decide to install, you must make a firm fixing for it in the ceiling – you can not just drive screws into the ceiling surface and hope for the best. What you do depends on the type of light you wish to fit.

Other fittings

Depending on their type, you may have to make provision for a special mounting method within the ceiling. Generally any fixture likely to weigh over 30-40 lbs should not be supported by the box alone. Many of today's ceiling lights are supplied with just a length of cord attached, and you have to connect this up to the circuit (and possibly the switch cable) using the correct size wire nuts. The National Electrical Code requires this connection to be made within an electrical box, that is set with its open end flush with the ceiling surface. The cables enter the box and are connected, the plate of the light fixture then covers the box when it is screwed into place. To fit the box you must cut a circular hole in the ceiling and fix a bearer between the joists just far enough above the ceiling to allow the box to be mounted with its rim flush with the underside of the ceiling surface. (Some boxes come complete with adjustable metal brackets that are screwed to the adjacent joists). Screw the box in place and, if necessary, repair around the rim using filler. Then feed in the circuit cables through the knock-out in the side of the box and make the connections to the fixture. You will need to make three connections if the fixture is in the middle of the run – two if it is at the end of the run. Make sure that you make all the ground connections before securing the fixture in place.

With some light fixtures the holes in the baseplate are at the same spacing as the threaded lugs at each side of the box, and you can mount the fittings with small machine screws driven into the lugs. Otherwise you must secure the fixture using an adaptor plate which is usually supplied with the fixture itself. Check the manufacturer's instructions supplied with the fixture for the exact method of installation.

WARNING

Always turn off the power at the service panel before beginning any work. Double check that the circuit is dead by means of a voltage tester.

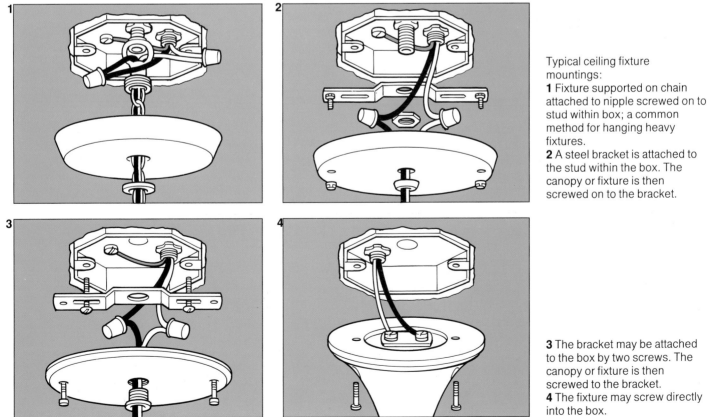

Typical ceiling fixture mountings:
1 Fixture supported on chain attached to nipple screwed on to stud within box; a common method for hanging heavy fixtures.
2 A steel bracket is attached to the stud within the box. The canopy or fixture is then screwed on to the bracket.

3 The bracket may be attached to the box by two screws. The canopy or fixture is then screwed to the bracket.
4 The fixture may screw directly into the box.

MOVING A PENDANT LIGHT

One of the simplest jobs you can do is to move a pendant light from one position to another – for example, so that the new position is precisely over the center of your dining table, or alternatively, to light up a breakfast bar or cooking range. On upper floors, you may wish to throw light specifically over a vanity unit or near a mirror in a dressing room.

Start by deciding on the precise position of the light, and make a hole in the ceiling at that point. If you do not mind leaving the old box in place (minus its pendant, of course) you can simply use it as a junction box, and run cable to the new light position, where you will have to fit a new box.

However, you will probably prefer to remove the box altogether and to repair the hole in the ceiling. In this case you need to fix a junction box above the ceiling at the site of the old fixture and run extra cable from there to the new fixture position. This is only possible if the junction box is to remain easily accessible – junction boxes, of course, must not be covered over permanently.

In the case of a plaster ceiling this is obviously not possible so the choice is to leave the box in its position and cover it with a special cover plate or to disconnect the wiring to the old box completely before plastering it over. In either case you will have to gain access to the existing wiring – by lifting first-floor floorboards to move a light in a ground-floor room or by gaining access to the attic above upstairs rooms. If you leave the old box in position, simply disconnect its fixture, feed in the prepared end of the new cable and connect its live and neutral wires to the same two wires that held the original fixture connections; splice the cable ground to the other ground wires. Then run the cable – parallel to or across the line of the joists, depending on how the floor is constructed – to the new fixture position and connect it up to the new box, making sure that it is compatible with the type of mounting bracket supplied with the new fixture. Attach the incoming ground wire to the box by means of a metal screw – except in the case of a chain-hung fixture. When installing a chain-hung fixture, run a separate number 18 grounding wire from the box to the fixture. Secure the mounting bracket in position before splicing the wires black to black and white to white. Finally screw the fixture securely in place.

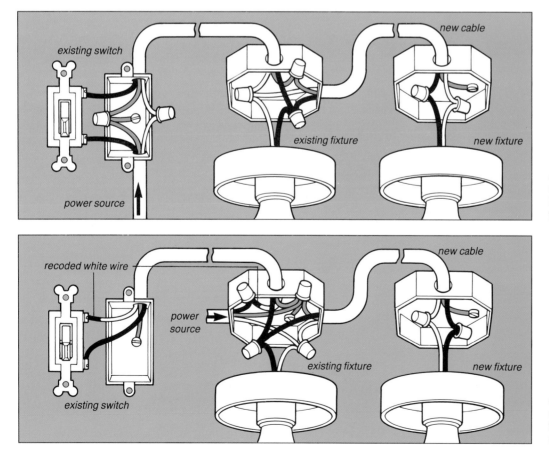

Addition of a new fixture to an existing end-of-the-run fixture. Both fixtures are operated by the existing switch.

Addition of a new fixture to an existing middle-of-the-run fixture. Both fixtures are operated by the existing switch.

Adding an extra light

If you want an additional light on the circuit, as opposed to just moving an existing one to a new position, you have to decide how to take the power supply from the existing circuit. You can either connect a branch cable into an existing fixture or cut the circuit cable at a convenient point and install a junction box.

In the first case, connect the black wire of the new cable under the wire nut containing the "hot" black wire and the neutral white to the other neutrals. Link the ground to the other ground wires.

In the second case, cut the circuit cable at a convenient point. Fit a 4 × 4 junction box, connect all the black wires together, connect all the white wires together and connect all the ground wires together. Run the new cable to either the switch box or the fixture box and connect as already described.

FITTING WALL LIGHTS

Installing wall lights involves little difference from adding ceiling fittings, except that you have to cut a cable chase (or fish the wire through the wall) and a recess for a mounting box in the wall surface. Wall lights are usually supplied with a short length of cord emerging from the body of the fitting, and so as with certain ceiling fixtures you have to connect it to the fixed wiring using wire nuts. You must house these in a box set in the wall as opposed to the ceiling.

Once you have decided on the position of the new light, mark round its baseplate on the wall surface. Drop a plumbline from the ceiling above it and mark the line of the cable chase. Draw around the box you intend to use, making sure that it is wholly within the baseplate outline, and cut out the chase and the box recess (see page 16). When the recess is deep enough, offer up the box and mark the positions of its fixing screws. Drill and plug the holes and then neatly fix the box in place.

Next, run in the cable drop to the light from the ceiling void, and run in the switch cable too if the light is to be controlled from a nearby switch. You can use round conduit to protect the cable if you wish or alternatively just plaster over it after securing it in the chase with cable clips.

Connect the light (and switch) cable(s) to the light as described (the fixture will be either middle-of-the-run or end-of-the-run) and tuck the wire nuts neatly into the mounting box. Finally fix the light baseplate to the wall to conceal the box.

On stud partition walls, using a stud finder to locate the studs, you can drop the cable down inside the partition (you will have to drill a hole in the head plate first, and you may have to cut away a little gypsum board to expose the horizontal bracing if any get in the way). If possible, try to mount the light close to the stud; so that the electrical box can be mounted on the stud to provide a firm fitting for the new fixture.

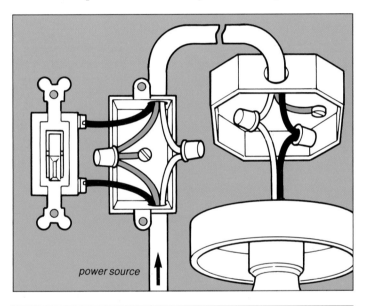

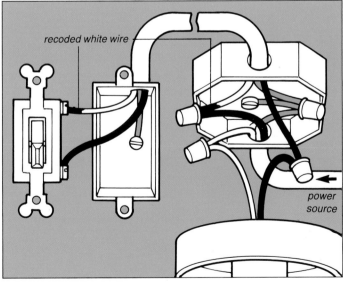

Top: New fixture and switch. End-of-the-run fixture, middle-of-the-run switch. For locating power source, see p. 45.

Above: New fixture and switch. Middle-of-the-run fixture, end-of-the-run switch. See p. 45 for locating power source.

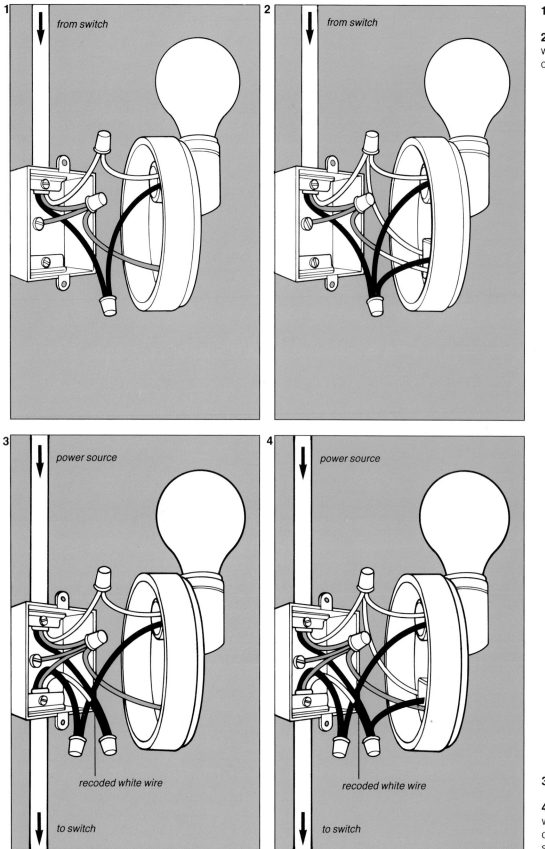

1 from switch

2 from switch

3 power source

recoded white wire

to switch

4 power source

recoded white wire

to switch

1 Middle-of-the-run wall fixture.

2 Middle-of-the-run wall fixture with integral outlet. Fixture controlled by the switch outlet.

3 End-of-the-run wall fixture.

4 End-of-the-run wall fixture with integral outlet. Fixture and outlet are controlled by the same switch.

PLANNING LIGHTING SCHEMES

Effective and attractive lighting schemes are even more difficult to plan than color schemes, and this probably explains why most homes appear to be distinctly unadventurous in this department. Yet this need not be the case: there is such a wide range of versatile and inexpensive light fixtures now on the market that anyone can afford to have a go at creating some unusual and highly decorative lighting effects in the home. To do so, it helps to know some of the basics about light outputs, light levels and the types of bulbs and fittings you can choose.

Outputs and light levels

Electric lamps (the proper term for light bulbs and tubes) are rated in watts(W). This measurement tells you how much electricity the lamp is using, but is not necessarily a very good indicator of how efficient the lamp is at producing light, and it certainly tells you nothing about how much useful light it produces or in what direction it emits it. The unit used for measuring light output is the lumen, and some lamps emit more lumens per watt (in other words, produce light more efficiently) than others. Amongst filament lamps, tungsten halogen are generally the most efficient, and incandescent lamps the least, and efficiency increases with lamp wattage. Fluorescent tubes are far more efficient emitters of light than filament lamps – on average, a tube gives out about four times as much light as a filament lamp of the same wattage.

The next thing you need to know is how much light to provide to light up a given room area. Obviously this depends on the tasks required in the particular area, whether the surface being illuminated is lit directly or indirectly, even the color of the surface, but as a rough guide for general lighting you need to provide about 2W per sq ft of room area. This is for minimum overall lighting; regard any additional local lighting in the room as being supplementary to this. Following the recommended lighting levels as set down by the Illumination Engineering Society of North America, lighting in a room should be determined firstly by what the room is to be used for followed by the appropriate type of lighting selected to suit the purpose.

Bear in mind that lighting can also be used to suggest spaciousness in a room – large luminous surfaces with contrast levels of lighted surfaces, for example, will appear bigger while lighting set to reveal surface textures will make a room appear smaller.

Downlights used to illuminate work surfaces and pick out plates hung on the wall.

A retractable fixture is useful for changing the mood of a room; used here over a dining table with a directional shade to good effect.

Light fixtures

Light fixtures come in five main types, and the characteristics of each will affect the quantity and direction of the light emitted (this will also depend on the type of bulb fitted – see page 38).

Pendant fixtures include simple pendant with one lamp-holder and multi-light units suspended from a ceiling-mounted rod and chain. They emit light either generally or in a direct/indirect and downward direction, depending on the type of shade fitted.

Surface-mounted fixtures are actually mounted on the ceiling surface, and so emit light only in a downward direction.

Wall-mounted fixtures (sconces) include very simple brackets (with or without shades) also spot-lights and uplighters (which can be free-standing).

Recessed fixtures are usually recessed into the ceiling void, and include downlights, adjustable fixtures and "wallwashers" – fixtures that cast a diffuse wash of light down a wall surface.

Freestanding lamps include the wide range of table- and floor-standing models offering general or local lighting according to type.

Light and shadow

Avoiding glare and hard shadows are two of the most important factors in lighting design, and answers are best given with reference to particular lighting requirements. For example, a center light in the living room is no good for reading (unless you sit right underneath it); what you need is a lamp behind you, casting light onto the page. In dining rooms you should avoid low fixtures where the naked bulb is close to eye level; use a rise-and-fall fixture over the dining table with a directional shade. In workrooms, you need a concealed bulb above your work surface to light up the work without dazzling, while in bathrooms concealed lighting above the mirror actually leaves your face in shadow; fixtures at each side of the mirror give better illumination.

Generally speaking, a good, home lighting scheme provides a balance of light where it is needed and soft shadow where it is not. Over-bright lighting is cheerless and severe, while dark shadows simply strain the eyes. In particular, fluorescent lights give a curious flat lighting effect without shadows – which may be fine for a kitchen, but not for the living room.

Always aim for an acceptable overall level of lighting, even when local lights (for reading, spotlighting a picture and so on) are providing extra illumination. The overall effect is then one of contrast, yet with no hard shadows.

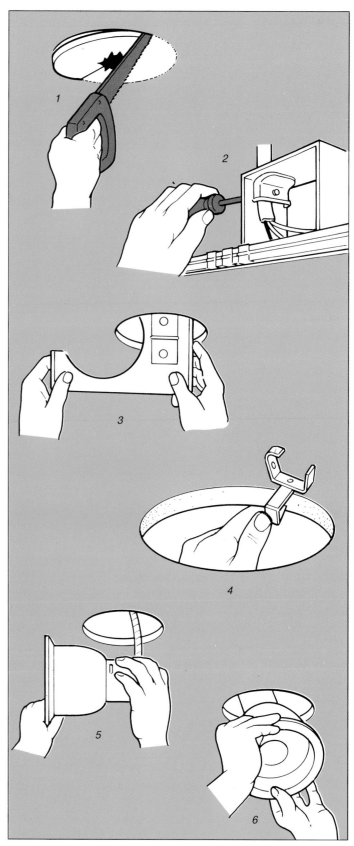

Downlights in an existing ceiling.
1 Cut the hole. 2 Connect the cable.
3 Insert the frame. 4 Clip into place.
5 Attach the trim. 6 Push into place.

FITTING DECORATIVE LIGHTS

Whether you wish to install recessed lighting, fluorescent strips, spotlight a picture or have a retractable lamp over your dining table, you will find the following information useful.

Downlights

Once the province of restaurants, hotels and offices, recessed light fixtures are now widely available for use in the home too, and one of the most popular types is the downlight. This is a can-like fixture that is installed within the ceiling void so that its bottom edge sits flush with the ceiling surface. Other alternatives include semi-recessed downlights, which are useful where the ceiling void is shallow, and also fixtures where the lamp is housed on its side – again taking up less vertical space.

Downlights are most usually fitted with internal silvered lamps which produce a fairly wide beam, but various other lamp types can be fitted too. Installation is comparatively simple. Once you have decided on the position of the fixture check that it does not coincide with a joist and that there is no pipework or electrical wiring immediately above it. Then mark a hole of the appropriate size on the ceiling with a pair of compasses, (many manufacturers supply a circular template of the exact size) and cut it out with a keyhole-saw. Where the ceiling is lath-and-plaster, reinforce the cut laths with battens pinned between the joists. Bring a single lighting circuit cable from a junction box through the opening and connect it to the downlight following the manufacturer's instructions. Fit the downlight into the opening and secure it by adjusting the clamps that attach on the upper, hidden surface of the ceiling.

Retractable fixtures

Rectractable fixtures are basically pendant lights with a spring-loaded wire that is drawn out or retracted to hold the light at any desired height below the ceiling. The cord is the curled type familiar on telephones, and the suspension wire is threaded through the center of the coiled cord. Their advantage is that they can be pulled down over dining tables, for example, during the meal, and pushed up afterwards.

Fitting such a lamp involves installing an octagonal box in a recess in the ceiling, screwed securely either to a joist or to a bearer fixed on battens between the joists. The power supply cable and the switch cable (if a switch loop is being used) enter the box and are linked to the wires from the fitting itself. Connect

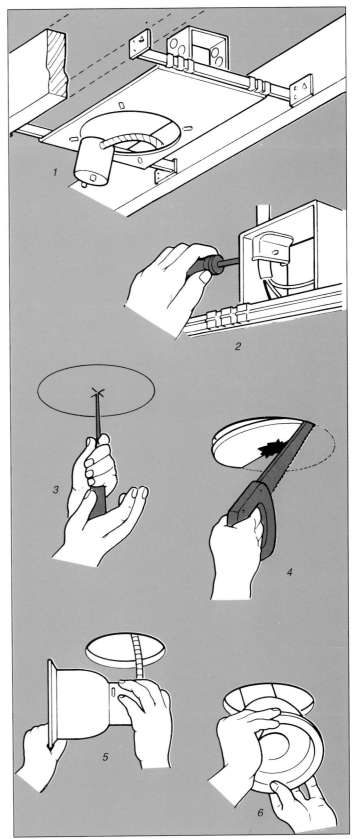

Downlights in a new ceiling.
1 Install the frame. 2 Connect the cable.
3 Install ceiling. 4 Cut the hole.
5 Attach the trim. 6 Push into place.

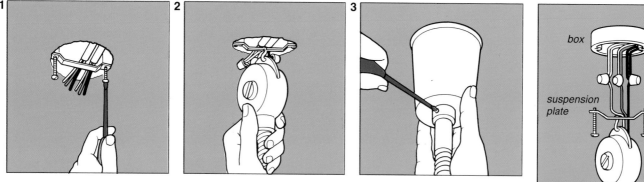

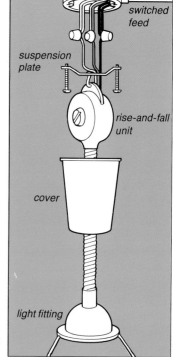

box

switched feed

suspension plate

rise-and-fall unit

cover

light fitting

black to black, white to white if no switch is being used. With a switch loop three connections are required. 1 Recode the outgoing white wire to the switch with a piece of electrical tape, making it essentially a black wire too. Join the recoded wire to the incoming black wire on the power cable. 2 Connect the returning black wire from the switch to the black wire on the fixture. 3 Connect the white wire on the fixture to the white wire on the incoming power cable. Remember to ground the fixture and the junction box.

Installing a retractable fixture:
1 Mount the hanging bracket onto the ceiling box.
2 Hook the fixture onto the mounting bar and make the cable connections.
3 Slide up the canopy and secure by means of a small screw.

Exploded view of the components of a retractable fixture.

Types of lamp

There is an enormous range of lamps and tubes you can use in your light fixtures. The most familiar are the everyday pear and mushroom-shaped opaque bulbs available in clear and colored glass and in a wide range of wattages. Decorative bulbs, widely used in wall lights, include candle lamps, decor round lamps in various sizes (intended to be seen through decorative shades) and pigmy lamps – small low-wattage bulbs often used in festoons. All these give off light in all directions unless the fitting itself modifies the beam.

Reflector lamps have a special inner coating which reflects the light in a particular direction. Internally silvered lamps are silvered round the base and sides and so give a broad beam in a "forward"-only direction. CS (crown silvered) lamps have the top of the lamp silvered to control glare, and rely on the fixture to direct the beam outwards. They produce a narrow beam suitable for spotlights. PAR (parabolic aluminized reflector) lamps have armored glass, give a broad beam and are intended for outdoor use.

Fluorescent tubes (circular, straight and U-shaped), pigmy lamps, PAR and reflector spots, GLS and decor lamps.

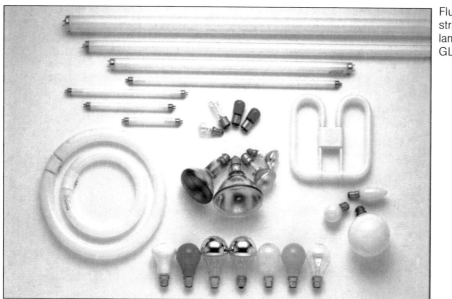

WARNING
Always turn off the power at the service panel before beginning any work. Double check that the circuit is dead by means of a voltage tester.

TRACK, BOWL AND WALLWASHERS

The following types of fitting will enable you to add variety to the lighting ambience in your home.

Track lighting

Track lighting is one of the most versatile types of lighting available, since it allows a range of different lighting effects from just one power source. The system consists of a length of special conducting track that is attached to ceiling or wall surfaces, and on which light heads of various types are mounted. The conductors within the track are safely shielded from touch, and each light picks up its power supply from special contacts once it is clipped into place. There is a wide range of heads – spots, floods and more diffuse beams can all be used from the same track. The only limitation is in the recommended maximum wattage the track manufacturer specifies, and the existing overall wattage on the lighting circuit you will use to power the track – remember that each 15 amp circuit can supply a maximum of 1800W. Various track lengths are available, and you can butt several lengths together if you want a continuous lighting "rail" across the room.

You can provide power to your lighting track either from an existing ceiling box – some types are designed so that all you do is connect the track into the box in just the same way as fixing an ordinary pendant light. Obviously, in the case of an existing ceiling box you will have to mount the track end over the box so that the wires can be connected up.

Once you have sorted out the power supply, you can start installing the track. Some systems use small clips that are screwed to the ceiling; with others you screw through the base of short stems that support the track. In either case, a firm fixing is necessary; either screw directly into joists (ideal if the track run coincides with a joist exactly) or fix bearers between the joists. With some track types you can drill fixing holes through the track to coincide with the centers of the joists.

Once the track is securely fixed in place, make the final electrical connection to your box, and clip the various light heads into place on the track.

Surface-mounted bowl lights

There is a wide range of surface-mounted bowl lights available, in round, square and oblong shapes with clear or translucent bowls in a great many styles and colors. They are useful where you want a diffuse light close to the ceiling – perhaps where headroom is limited, such as at the foot of the stairs, and are obligatory in bathrooms.

Most are mounted in a similar way, with a fixing bar being screwed to the outlet box first and then a baseplate being secured to this. If you are fitting one in place of an existing ceiling fixture, you simply have to remove the old fixture; there is generally no need for a new box. Screw the new bar to the box that supported the old fixture, then connect the existing circuit cable to the terminal block on the baseplate fitting. You can then attach the baseplate (which carries the lampholder) to the fixing bar, and replace the bowl after fitting a bulb of the recommended type and wattage. Some are held in place with small screws or spring-loaded clips, while others have a sliding level that locks the bowl in place.

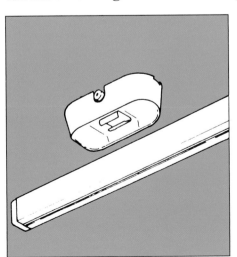

Methods of fixing lighting track to a ceiling: surface-mounted clips connected by cable to a nearby ceiling junction box.

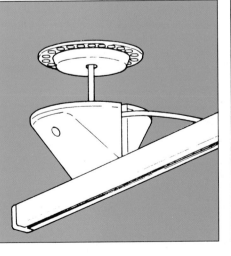

Mounting canopy fitted over an existing ceiling feed; the canopy also conceals the slack flex necessary for making connections.

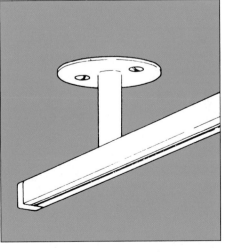

Mounting stems screw directly to a BESA box; both this and the surface-mounted clip can be used for fixing track to a wall also.

Adjustable downlights and wallwashers

Apart from straightforward downlights, there are several variations on the recessed lighting theme that can perform more varied lighting functions. The downlight simply casts a uniform beam straight down at the ground beneath the fixture – a virtual pool of light. Wallwashers are basically directional downlights, and are usually installed 2 to 4 feet from a wall of the room so that the light from the fixture literally "washes" the wall with light. The fixture is adjustable, allowing the angle of incidence to be varied to achieve precisely the desired effect. Several wallwashers installed in a line can be used to wash an entire wall with light, and if colored lamps are used very attractive and dramatic color scheming effects can be achieved – literally changing the day-time color schemes of a room at the flick of a switch.

Recessed adjustable downlight fixtures are another type of downlight, and literally work like an eye – the lamp itself is housed in a spherical mounting which can swivel within the main body of the fixture to almost any angle. If fitted with a spotlight it can be used to highlight individual features in the room in a far less obtrusive manner than an ordinary ceiling- or track-mounted spotlight.

Installation and electrical connections are much the same as for a downlight – power for the light can be taken as a spur from a nearby junction box unless the fitting is replacing one already in that position.

> **WARNING**
> Remember to use only materials and fixtures that are clearly labeled with a U.L. mark.

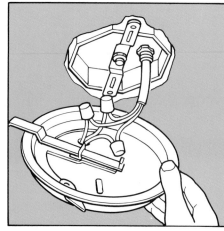

Screw the mounting bar into the outlet box and make the wire connections.

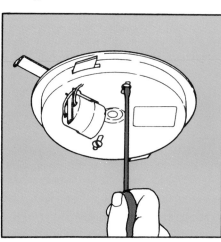

Screw the fixture base to the mounting bar.

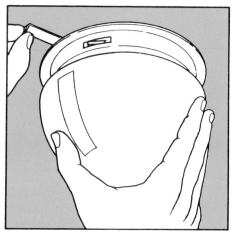

Hold the bowl in place and secure by pushing in the lever or by tightening the supporting screws.

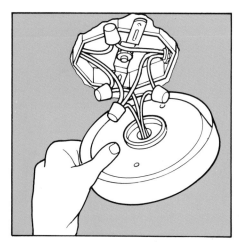

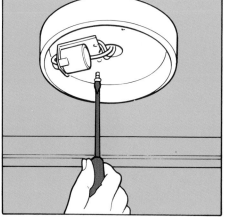

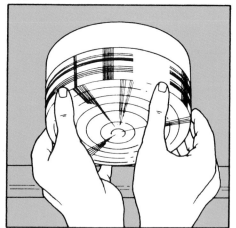

Follow the first two steps described above, but then secure the bowl by twisting firmly and securely into position.

INSTALLING A FLUORESCENT LIGHT

Fluorescent lights are undoubtedly more efficient than filament lamps in terms of light output per watt of power consumed, but the light they emit is usually felt to be rather flat and unsympathetic to the eye. However, they are excellent task lights for rooms like kitchens, bathrooms and workshops, and can be very effective when used as concealed lighting – behind a valence or baffle, for example, casting light downwards over drawn curtains or a kitchen worktop, or flush-mounted behind opaque diffusers in a suspended ceiling.

Most fluorescent lights are linear tubes, ranging in length from 6in up to 8ft and in diameter through three main sizes – ⅝in (miniature), 1in (slimline), 1½in (regular). Domestic sizes are 4ft producing a light equivalent to 40W and 6ft similarly with an output of between 65 and 80W. Circular tubes are also made, commonly in diameters of 12 and 16in. All types are made in several "colors" – ranging from a cold white through to "warm white" – the best choice for most domestic use.

The simplest way of installing a fluorescent light is to buy a complete unit containing all the controls required – starter and choke or ballast – housed within an enclosure that is fixed to the ceiling. These are available for both linear and circular tubes; a range of diffusers is available.

To install the fixture, start by using the template provided (or the backplate of the fixture) to mark the fixing holes on the ceiling. Ideally the fixture should be screwed directly to a joist (or to several if its position is at right angles to the joist line), but if this is not convenient you will have to fix braces between the joists. Screw the baseplate in position, feeding the circuit cable(s) and connecting them. If your lighting circuit has only live and neutral wires, run in a separate ground wire from the terminal block back to the main grounding point at the service panel.

Next, fit the backplate cover in place and add the spring-loaded clips at each end which hold the tube in place. Slot the tube into place, fit the starter in the side of the baseplate and test that the light works.

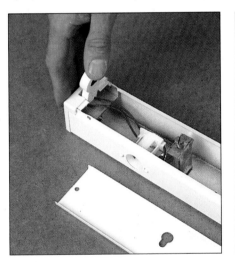

Fitting the end connectors to the ends of the fluorescent tube base; they may be fixed rigidly or spring-loaded.

After locating a ceiling joist, marking through the fixing holes in the base of the fitting on to the ceiling.

Connecting the cable cores to the terminal block; make sure it is firmly clamped to the metal lug on the base.

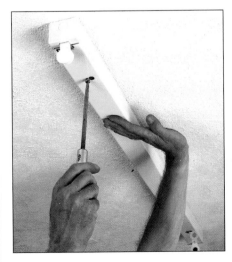

Securing the cover plate; this may have double fixing holes, one the size of the screw head which it fits over.

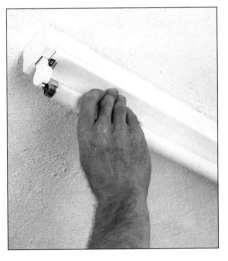

Fitting the tube into the end connectors by pushing in and twisting through 90°, or by springing out the caps.

Inserting the starter switch by pushing it into its holder until it locates in the sockets, then turning clockwise.

CHAPTER 4
ADDING LIGHTING AND POWER

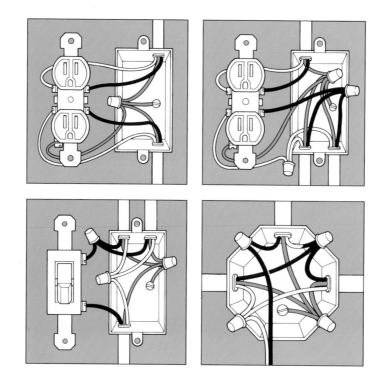

To get the maximum use from your existing wiring system, extra spurs can be added to both the lighting and power circuits provided that they have the necessary reserves. Single outlets can be changed into doubles which will ease the problem of having to use multiple adaptors with their attendant trailing cables. But where existing circuits have no reserves and cannot be further extended, you will need to add a completely new circuit or subpanel. The following chapter explains in detail how to do this, and finally, shows you how to fix outlets in walls so that you will be able to complete the work confidently and professionally.

ADDING LIGHT AND POWER

If you have added an extra room to your house – as an extension or an attic conversion – you will naturally want to supply it with lighting and power circuits, perhaps even with special circuits as well if it is a kitchen or bathroom and you want to install equipment like an electric clothes washer and dryer.

The first thing you need to establish is exactly what your requirements are. Use a copy of the plans for the new room to mark in where you want light fixtures, switches, outlets and other special outlets. Then work out if it is possible to extend the existing lighting and power circuits to supply the new room, or whether new circuits will be required.

In the case of existing circuits, each 15 amp circuit has a capacity of up to 1800 watts and each 20 amp circuit has a capacity of up to 2400 watts, so you need to check how much demand is on each circuit already, remembering to leave some reserve – see Extending Lighting Circuits on page 29 for more details of exactly what is involved. If there is scope to extend the circuit, you can then work out the best point at which you would make a connection into it – at an existing junction box or at a new junction box cut into the circuit cable at a convenient point. Mark this on your plan, and then sketch in the likely cable routes to light fixtures and their switches, so you can estimate how much cable will be needed for this part of the job. Do not forget to include three-wire and ground cable too if you plan to use any three-way switching arrangements. If your existing circuits already supply their quota of outlets, you will have to install a new circuit. The obvious way of doing this is to use a spare fuseway at the service panel (if you have one) or by means of a new service panel (if you have not) – see pages 51–52 for more details. Each 15 amp circuit can serve an area of up to 375 sq feet, each 20 amp circuit can serve an area of up to 500 sq feet.

If the extra area of your new room(s) can be added to the area served by an existing circuit without exceeding these area figures, then you can extend this circuit as long as it has some reserve capacity. You can connect the new cable to it either from the last outlet on the circuit or at any convenient point along it. In the latter instance a connection can be by means of a junction box cut in at a convenient point.

You can extend a circuit in one of two ways – by adding spurs to the circuit, or by extending from the end but do not make it longer than 75 feet. You can add as many spurs to a circuit as there are outlets on the original circuit, and the actual connection can be by outlets or junction boxes cut into it. If your new rooms adjoin existing ones, you can run spurs through the wall from the existing room to the newly built one.

Any other special circuits your new rooms need will have to be wired in as completely new circuits from the service panel.

SINGLE OUTLETS TO DOUBLE

An obvious way of increasing the number of outlets you have available without having to interfere with the fixed wiring in any way is to remove them and fit two outlets in their place or even three or four!

In principle, all you have to do is to turn off the power, remove the fuse protecting the outlet, unscrew the existing single outlet and attach another one alongside. The obvious complication is with the mounting box; that will have to be changed and here you have several options.

Surface-to-surface conversion

If your existing single outlet is surface-mounted and you want the new one to be mounted in the same way, buy a new double outlet with matching surface box, then unscrew and disconnect the old cover plate from the outlet, and unscrew the old single mounting box from the wall. Lift is away, leaving the circuit cable protruding from the wall (or from raceway alongside if this has been used). Remove an appropriate knockout from the new double box, feed in the cables and put the box against the wall to mark the positions of the fixing screws. Drill and plug the holes and fix the box in place, checking with a spirit level to

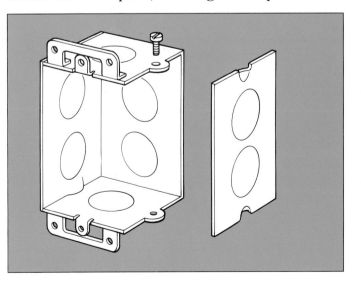

The side of the metal outlet box can be removed in order to join several boxes together.

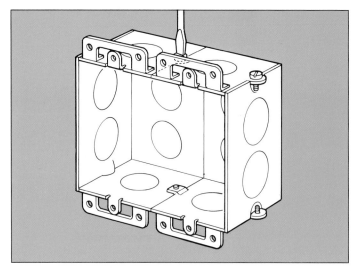

A pair of boxes joined to make a double box.

A pair of receptacles wired in the middle of the run.

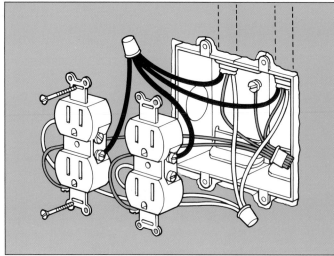

make sure that it is absolutely level. Finally reconnect the wires to the terminals on the outlet and screw it securely to the box.

Flush-to-flush conversion

If the old single outlet is flush-mounted and you want the new one to be flush as well, you have a little more work to do. If the wall is solid masonry, begin, as before, by removing the original cover plate. Then unscrew the fixings holding the single box in its recess, and prise it out – you may need to cut around the sides of the box to free it, using an old knife. Now check how much play you have on the circuit cables; if there is very little, you will have to enlarge the recess at each side, but if there is a fair amount of slack you can cut away the brickwork at one side only – the cable should still be able to reach the new terminals.

Place the new double box against the wall in the required position and mark its outline over the existing hole. Within the outline, drill a series of holes in the brickwork to the same depth as the box using a depth stop on your drill, and then cut away the honeycombed masonry with a sharp mason's chisel and a light sledge hammer. When the hole reaches the right size and depth, test the box for fit, and then mark the position of the fixing screws. Remove the box, drill and plug the holes and fix the box securely in place after feeding in the circuit cables. Finish round the edges of the box with filler, reconnect the circuit cables to the new receptacles and screw them into the box, and finally install a double cover plate.

Where you plan to carry out the job on a stud partition wall, exactly what you do depends on what you find when you remove the old single cover plate. If the box is screwed to a batten fixed between the studs, simply remove it, enlarge the hole in the wall to match a double box and screw this in place to the batten. Feed in the circuit cable and reconnect as before. If the box was secured with metal clips, you could re-use them to fix the new double box. However, as already explained, this is a fiddly job and does not give a very secure fixing; it is much better in this situation to discard the clips and to fit a double cavity mounting box instead.

WARNING

Always turn off the power at the service panel before beginning any work. Double check that the circuit is dead by means of a voltage tester.

ADDING A RECEPTACLE

As we have seen there are limitations placed on the number of electrical appliances and fixture we can use on any one circuit, based on the cable size and the fuse or circuit breaker protection offered. Generally a 20 amp circuit of number 12 wire should take care of about 500 sq ft of floor area, or a 15 amp circuit 375 sq ft. However, to offer some reserve in the case of circuit failure and to help distribute some of the load, two or more circuits will supply outlets in any room.

Almost all of the outlets and light fixtures on these circuits could serve as a power source for the installation of a new electrical outlet. What matters most is that you select a power source that is convenient to the position of your new receptacle and one that is on a circuit with enough reserve capacity to suit your needs. (see p 29).

Selecting a power source

The following are the three main options possible.

1 End-of-the-run receptacle This is the simplest location to make a power hook-up. Only one cable comes into the box and a black and white wire are connected to the receptacle, while two of the terminal screws remain vacant. To tap off power bring in a new cable. Attach its black wire to the vacant brass screw and then attach the neutral white wire to the vacant silver screw. Finally splice together the ground wires.

2 Middle-of-the-run outlet In this case two cables enter the box and all four terminal screws on the receptacle are occupied. To tap off power remove one pair of black and white wires from the terminals on the receptacle and replace with two short jumper wires about 4 inches long; a black jumper wire on the brass screw and a white jumper wire on the silver screw. Bring in the new cable, join the two black wires and the black jumper wire with a wire nut, join the two white wires and the white jumper wire with a wire nut. Finally splice together the ground wires.

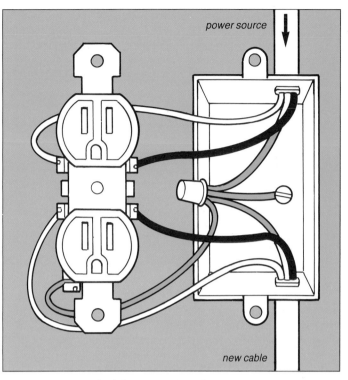

Tapping into an end-of-the-run receptacle.

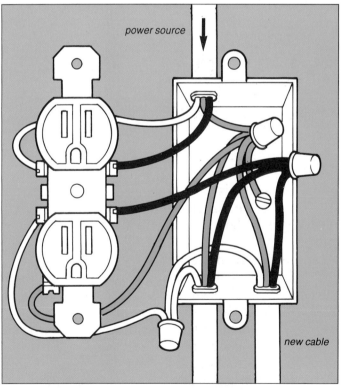

Tapping into a middle-of-the-run receptacle.

WARNING

Always turn off the power at the service panel before beginning any work. Double check that the circuit is dead by means of a voltage tester.

3 Middle-of-the-run switch In this case two cables enter the box; the white wires are spliced together and the two black wires go to the switch terminal. It is then necessary to determine which of the two black wires is the power feed. This must be done by use of a voltage tester and entails switching the power on again after the switch has been removed. Remove the black wires from the switch terminals, have an assistant return the power, then using a voltage tester touch one of the tester probes against the metal junction box. Use the other probe to touch the two black wires in turn – the black wire that lights the tester is the power source. Turn off the circuit before continuing with the operation.

Once you have identified the "hot" black wire reattach the other black wire to the switch terminal. Cut a short black jumper wire about 4 inches long and attach this to the remaining switch terminal. Bring in your new cable, splice together the three white neutral wires with a wire nut. Splice together the two black wires and the black jumper wire from the switch with a wire nut. Finally splice together the ground wires to complete the job.

Middle-of-the-run fixture

Again it is necessary to identify the "hot" wire by the use of a voltage tester. Several cables may enter the box and there will be at least two splices containing black wires. The black wire coming from the fixture itself will be connected into one splice – this splice is unlikely to contain the live wire.

In order to identify the live wire in the other splice, turn off the power to that circuit, remove the wire nut and separate out the black wires, making sure that they do not touch anything. Have an assistant return the power, then using a voltage tester touch one of the probes against the grounded junction box, touch the other probe to the black wires in turn – the one that lights the tester is the live wire from the service panel. Switch off the power before continuing. Bring in the new cable and splice the black wire together with all the black wires that were under the wire nut containing the live wire. Add the new white wire to the splice containing all the white wires. Finally splice together all the ground wires.

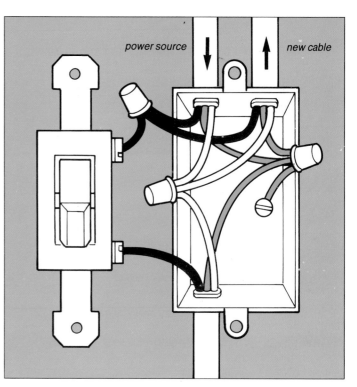

Taking power from a middle-of-the-run switch.

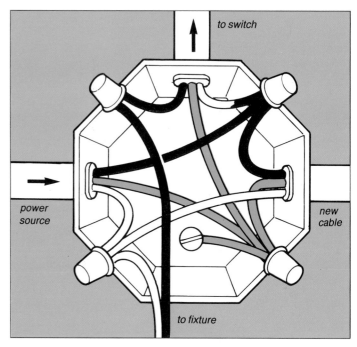

Taking power from a middle-of-the-run fixture.

WARNING
Always turn off the power at the service panel before beginning any work. Double check that the circuit is dead by means of a voltage tester.

FIXING OUTLET BOXES TO WALLS

The two tricky parts to extending your power circuits, once you have decided where the extra outlets are going to be are in running in the new cables (dealt with on page 16) and mounting the new outlets on the wall. Exactly how you go about the second part of this job depends on whether you are installing surface-mounted or flush outlets, and whether you have walls of solid masonry or of stud partition construction.

Surface-mounted outlets

Mounting your new outlet on the wall surface is certainly the easier option, whether you are running the new cables on the surface (clipped in place or run in raceway) or hiding them beneath plaster and floorboards. The outlet faceplate is in this case screwed to a plastic mounting box (usually about 1⅜in deep) which is itself screwed to the wall. The cable is fed into the box from the side if it is run on the wall surface, and from the back if it is concealed.

Flush outlets

Mounting outlet boxes flush with the wall surface is the neater solution. There is less risk of the outlet being damaged accidentally, and it is certainly less obtrusive. Rather more work is involved in fitting an outlet in this way, since the mounting box – in this case of galvanised metal – has to be fixed in a recess cut in the wall.

The cable enters the box through a hole knocked out of the side or base of the box. Then the outlet and its faceplate are connected up and screwed to the box to complete the job; now some adjustment is possible because the receptacle mounting screws are positioned in slots which allow the box to be positioned vertically.

With hollow walls under construction the box is usually mounted on a batten fixed between the studs, and a hole is cut in the drywall over it. With existing walls it is possible to mount the box in a cut-out in the drywall by using small lugs clipped to the box sides, but it is an extremely fiddly job and the fixing is not particularly secure. A better and easier solution is to use a plastic cavity mounting box, which has a flanged edge and two spring-loaded clips that grip the back of the drywall or paneling when the box is pushed into place within the cut-out. The cables are fed into the box through a knockout before the box is positioned.

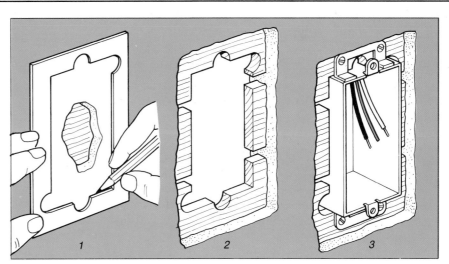

To flush mount a box in a lath and plaster wall:
1 Mark the box outline. 2 Cut the hole. 3 Fish the wires through and connect into the box, then mount the box and secure to the lath with screws.

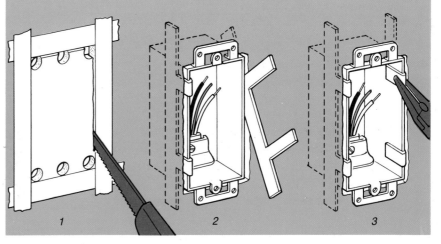

To flush mount a box in a gypsumboard wall:
1 Mark the box outline and cut out the hole. 2 Fish the wires through and connect into the box; push the box into position. 3 Bend the clips into position with a pair of pliers.

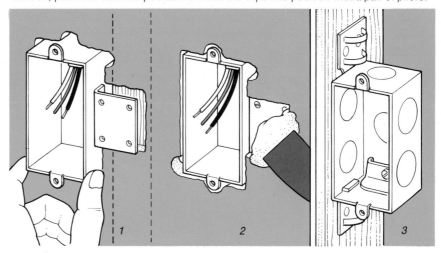

If the box is to be mounted close to a stud, the box shown above can be nailed to the stud by means of a metal flange. In new work (as shown on the right), the box can be nailed directly to the side of the stud before the drywall is installed.

CHAPTER 5
ADDING TO THE SYSTEM

Making extensions to your existing power circuit will
help greatly to alleviate an inadequate wiring system, but in
certain circumstances, such as installing a high-rated appliance,
like a dishwasher for example, there is no other alternative
but to add a new circuit.
This chapter tells you how, and also which portable domestic
appliances can safely be plugged into an existing circuit and
lists the larger appliances that should be wired to their own
circuits — along with typical running costs for a wide variety of
appliances. Finally, information is also given on installing a
domestic alarm system to keep your well-equipped home
sound and safe from intruders.

EXTENDING POWER CIRCUITS

Improvements to your power circuits are generally far more of a necessity than those to your lighting arrangements. Less-than-perfect lighting may be a nuisance and a disappointment, but inadequate provision for your electrical appliances can actually lead to situations that are dangerous.

If you do not have enough electrical outlets in the right places, you are likely to make extensive use of potentially overloaded adaptors and extension cords plugged into the few outlets that are available. The end result is an increased risk of electrical accidents caused by short circuits, poor contacts between plugs and outlets leading to overheating, and overloading of the circuits causing persistent fuse-blowing and the risk of fire.

As a guide to how many outlets a well-equipped home should have, the Electrical Code requires that electrical outlets should be placed no more than 12ft apart, but as a general guide outlets can be placed 10ft apart with a minimum of three per room and kitchen outlets placed about 4ft apart.

You may well find that you need more outlets than this in some rooms and fewer in others; the best way to find out is to make a room-by-room plan of the house and mark in what appliances you keep in a particular position (and which therefore need an outlet full-time) and which appliances you tend to take out and use occasionally (these can share an outlet). Mark the positions of existing outlets, noting whether they are singles or doubles and then add on the outlets you require. You may need just one or two extra outlets, which can probably be added to the existing wiring or you may require so many that new circuits will be called for.

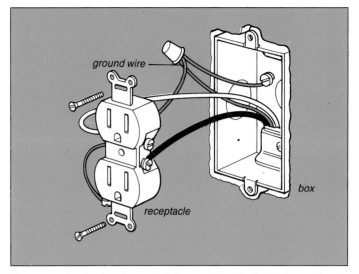

A receptacle wired at the end of the run. The black wire is attached to the brass screw and the white wire to the silver screw.

What is there already?

Once you have worked out what your requirements are, you must examine your existing wiring to find out where the circuits run, what type they are and what condition they are in.

To take the first point. In order to establish which circuit your various outlets are on, turn off the power on the main switch, remove one power circuit fuse and plug a working appliance into each outlet in turn. Mark on your plan those that are controlled by fuse 1, and repeat the process for as many power circuit fuses or breakers as in your service panel.

Lastly, you must check up on the condition of the system, to see whether it is safe to extend it. The cables are the most important factor here; inspect them at the service panel and where they run into individual outlets. Check as many as you can, with the power switched off for safety, in case parts of the

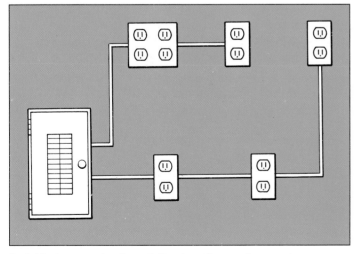

Individual power circuits radiating from the service panel.

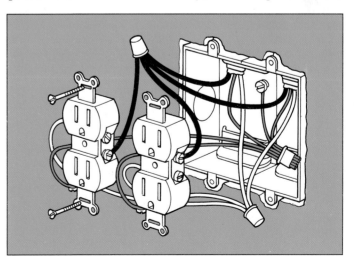

A pair of receptacles wired in the middle of the run.

system have been extended since the original installation. What you are on the lookout for is woven fabric sheathed cable. This has been obsolete for years, so if you find this type of cable, you should seriously consider completely rewiring the circuits concerned. Not to do so may invite the risk of fire. If in doubt call in a licensed electrician to check the wiring.

If you find that your circuits are well wired in modern plastic sheathed cable and feed modern outlets, then you can probably carry out all the extension work you require with the minimum rewiring. However, you need to understand the principles behind each type of circuit, to understand the limitations placed on each one by the requirements of your local electrical code, to which all new wiring work should conform.

WARNING

Make sure that you are familiar with the type of wiring existing in your house before attempting any work. If you are at all in doubt consult a licensed electrician or your local Building Inspector.

DOMESTIC APPLIANCES

Portable domestic appliances with ratings of up to 1500W, such as those items listed in the box below, can be plugged in to an existing ring or radial circuit using a regular 20 amp receptacle. However, it is important to remember to remove the plug from the receptacle when the appliance is not in use, since there is always the risk of fire, particularly if the circuit ever became overloaded. Because these are the plugs that are in constant daily use, check them regularly to make sure the wires are firmly attached.

Appliances	Watts	Voltage	Wire size	Fuse/ breaker size
Blender	300	120	12	20 amp
Can opener	150	120	12	20 amp
Coffee maker	700	120	12	20 amp
Deep-fat fryer	1500	120	12	20 amp
Dehumidifyer	300	120	12	20 amp
Fan	300	120	12	20 amp
Fryer	1200	120	12	20 amp
Iron	1000	120	12	20 amp
Floor lamp	300	120	12	20 amp
Food mixer	150	120	12	20 amp
Radio	100	120	12	20 amp
Room heater	1500	120	12	20 amp
Sewng machine	100	120	12	20 amp
Stereo	300	120	12	20 amp
TV	300	120	12	20 amp
Vacuum cleaner	600	120	12	20 amp

Installing domestic appliances

Larger appliances, such as ranges and water heaters that draw heavier amounts of current should be wired to their own circuit. The wiring will be heavier than the normal house wiring and a special outlet is usually required. A range of outlets rated from 20 to 60 amps is available. The arrangement of the slots and the shapes of the individual prongs vary according to the amperage, which is clearly marked on the front of the receptacle and should be carefully checked before use. However, manufacturers of heavy household appliances supply the appliance with a power cord and plug to match the relevant outlet and, provided these are used, there is no risk of overloading. Note the manufacturer's instructions, and never substitute a different plug or outlet for the one supplied.

Appliance	Watts	Voltage	Wire size	Fuse/ breaker size
Air conditioner	1200	120/240	12	20 amp
Clothes washer	700	120	12	20 amp
Dish washer	1200	120/240	12	20 amp
Dryer	5000	120/240	10	30 amp
Freezer	350–500	120	12	20 amp
Furnace or boiler	800	120	12	20 amp
Garbage disposer	300	120	12	20 amp
Range with oven	1200	250	6	50–60 amp
Range top (separate)	5000	120/240	10	30 amp
Range oven (separate)	5000	120/240	10	30 amp
Refrigerator	300–500	120	12	20 amp
Water heater	2000–5000	120	10	30 amp

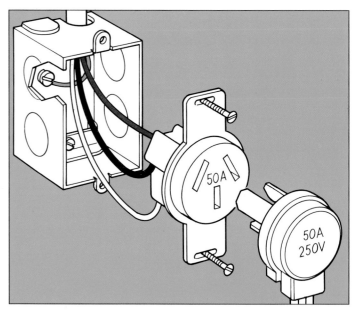

A heavy-duty appliance outlet. The receptacle will only accept a cord and plug of the correct rating; both are clearly marked.

Running costs of domestic appliances

The cost of running electrical appliances in your home is based on the number of units of electricity they have used over a given period. Each unit represents the amount used in one hour by a 1kW appliance. An appliance rated at 3kW, for example, uses the same amount of energy in twenty minutes.

The table below shows you how much electricity is used, on average, by everyday household appliances with different kW ratings. A 100W light bulb, for instance, will give 10 hours of light before it uses one kilowatt unit, whereas a 3 kilowatt electric fire will produce heat for 20 minutes only for the same 1 kilowatt.

TYPICAL RUNNING COSTS

Appliance	Typical use	No. of units	Appliance	Typical use	No. of units
Range	Meals cooked for 4 people for one day	2½	Vacuum cleaner	In use for 1½-2 hours	1
Deep-fat fryer	In use for 1 hour	1½	Range hood	24 hours continuous running	2
Microwave	Cooks two average size joints of meat	1	Extractor fan	24 hours continuous running	1
Storage heater	Provides heating for one day	8½	Tin opener	6000 tins	1
Fan heater (3kW)	Heat for one day	3	Clock	Continuous running for 1 week	1
Instant water heater	Heats 2-3 bowls of water	1	Single under-blanket	Warms the bed for one week	1
Immersion heater	Hot water for 4 people for one day	9	Single over-blanket	Warms the bed for one week	2
Instant shower	1-2 showers	1	Hair dryer	2 hours running	1
Dehumidifyer	24 hours continuous running	3	Shaver	Gives 800 shaves	1
Dishwasher	Washes one full load	2	Power drill	In use for 4 hours	1
Automatic clothes washer	Washes one full load	2½	Car battery charger	In use for 24 hours	1
Clothes dryer	One full load	2½	Hedge trimmer	In use for 2½ hours	1
Refrigerator (4 cu ft)	Keeps food fresh for 1 week	7	Cylinder lawn mower	In use for 3 hours	1
Freezer (6 cu ft)	Keeps required temperature for one week	9	Hover mower	In use for 1 hour	1
Towel rail	Continuous heat for 6 hours	1½	Stereo system	Plays for 8 hours	1
Coffee percolator	Makes 75 cups	1	Color television	Gives 6 hours viewing	1
Foodblender	Blends 500 pints	1	VCR	Records for 10 hours	1
Food mixer	Mixes 60 cakes	1	Sewing machine	Sews for 24 hours	1
Iron	In use for 2 hours	1	100 W light bulb	Gives 10 hours light	1
			40W fluorescent strip	Gives 20 hours light	1

ADDING A NEW CIRCUIT

Extensions and additions to your existing lighting and power circuits may go a long way towards alleviating the problems of an inadequate electrical system, but sometimes there are circumstances where you have no alternative but to add a new circuit. A common instance is where you want to install a high-rated electrical appliance such as a range, that cannot be run off an existing power circuit. Another is where you want to upgrade your lighting system and your existing circuits have their full quota of lighting points.

The simplest way of doing this is to make use of an existing spare fuseway in your service panel. You will be unlikely to find a spare in any but a recent electrical installation – most electricians are loath for obvious financial reasons to install a bigger panel than the installation requires, but this is very often false economy. If you are having rewiring work done that involves fitting a new service panel, choose a size that allows one, two or even three spare fuseways so that you can, at any time in the future, extend the system with the minimum of upheaval.

What type of circuit

The new circuit, whether for lighting or to feed an individual appliance, will be wired up as a radial circuit. If it is for extra lighting, it should be run in number 12–2 wire in Romex or BX (use what is already installed in your home. If you have BX use BX cable, if it is Romex use Romex cable) and the circuit fuse or breaker should be rated at 15 or 20 amps.

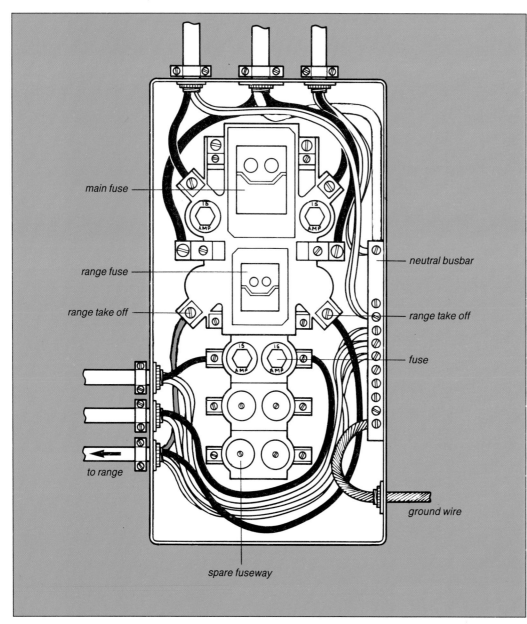

main fuse

range fuse

range take off

to range

spare fuseway

neutral busbar

range take off

fuse

ground wire

Typical fuse panel with a pull-out block for the main fuse and for the range. The main cartridge fuses are mounted on one or two non-metallic pull-out blocks. By pulling firmly on the hand grips, you can remove the blocks from the cabinet and disconnect all power.

Connections at the service panel

It is customary to arrange the circuit fuses in modern service panels on the live busbars in such a way that the highest-rated fuse or circuit breaker is nearest to the main isolating switch and the lowest-rated one is furthest away. You may therefore have to alter the position of some of the existing breakers to allow the new one to be placed in the right position. In modern service panels it is easy to unclip individual breakers and simply move them along the busbar after turning off the isolating switch on the main service panel. The new circuit breaker is then clipped into place on the busbar, ready to receive the new circuit cable.

In the diagram below, the main disconnect switch is at the top of the panel and spare breaker at the bottom – the furthest point from the main breaker.

The circuit cable

Strip enough from the cable so that the wires will reach the most distant connection point – (usually the neutral busbar); fir a two point clamp and feed the cable into the panel through a convenient knockout, cutting overlong wires to the correct length if necessary. Connect the live wire to the terminal on the fuseholder or circuit breaker, the neutral wire to the neutral block and the ground to the main grounding terminal, usually the neutral busbar. Check that all the connections are secure and correctly made.

The last stage is to refit the protective cover over the panel and to turn on the main isolating switch again so you can test the new circuit.

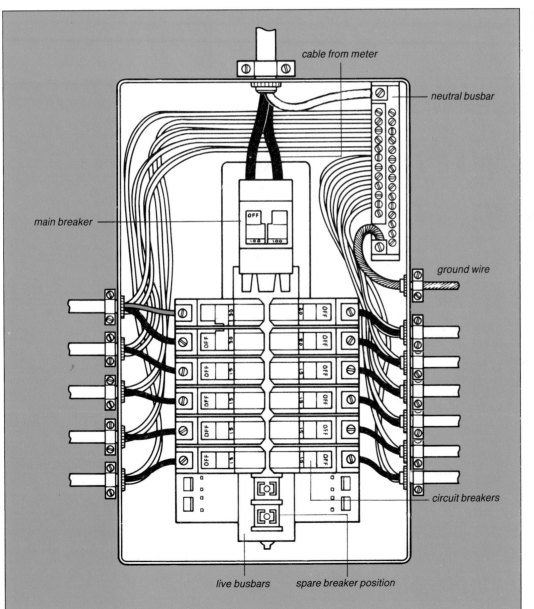

Circuit breaker panel with a main disconnect at the top.

cable from meter

neutral busbar

main breaker

ground wire

circuit breakers

live busbars   spare breaker position

WARNING

Always turn off the power at the service panel before beginning any work. Double check that the circuit is dead by means of a voltage tester.

ALARM SYSTEMS

Although an alarm may be considered a luxury, the ready availability of systems designed for installation at reasonable cost makes this form of home protection well worth considering.

Usually alarm systems come in the form of a basic kit which can often be extended and modified with accessory packs to meet your needs and property exactly. Most are of the perimeter type; that is they rely on sensors detecting someone breaking in through a door or window to trigger the alarm. Some have sensors that detect movement inside the house.

In addition to door and window sensors – usually in the form of magnetic switches – an alarm kit will include the control unit, alarm sounder (a bell or siren) and the necessary wiring. Some kits may even provide mounting screws, wall plugs and cable clips as well. If movement detectors are offered, they usually take the form of pressure mats which can be laid under the carpet in doorways, in front of the TV set or at the foot of the stairs.

The system will operate from the mains supply but a good kit will also provide a standby battery in the control unit in case the mains power is cut off. The sounder should also have its own battery so that it will operate if the wiring between it and the control unit is either accidentally or deliberately cut.

Sensors are screwed to wooden window frames or stuck to metal ones with self-adhesive pads, then the wires are run back unobtrusively to the control unit (mounted out of the way but still within easy reach). The sounder must be mounted high on a wall outside where it is plainly visible.

Automatic light switches

Unlit houses at night can advertise the fact that no one is at home and is just what a potential burglar wants to know. However, you can give the impression that someone is in with automatic light switches. These are designed to replace regular light switches and can be programmed to switch lights on and off at preset times every night or in a random fashion to give a more realistic impression.

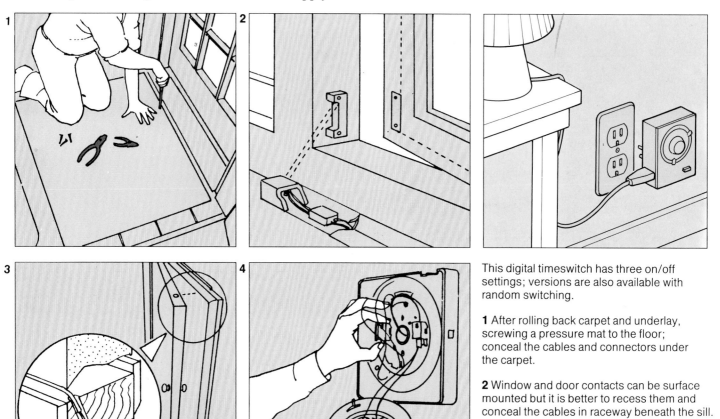

This digital timeswitch has three on/off settings; versions are also available with random switching.

1 After rolling back carpet and underlay, screwing a pressure mat to the floor; conceal the cables and connectors under the carpet.

2 Window and door contacts can be surface mounted but it is better to recess them and conceal the cables in raceway beneath the sill.

3 Flush-mounted versions should be positioned so that there is a gap of no more than ¼in between them and the magnet.

4 Connecting the circuit cable to the bell high on a wall; seal behind it with mastic.

CHAPTER 6
WIRING OUTDOORS

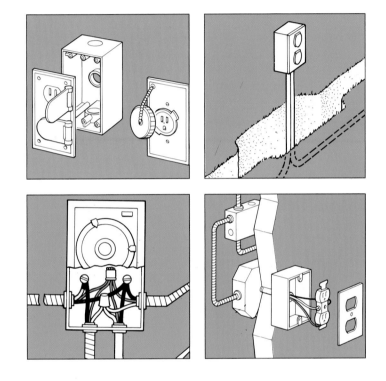

You may wish to add a whole new lighting system to illuminate your barbecue area or to enhance your outside landscaping, or simply to light up your front door and walkways. This chapter illustrates such situations emphasizing the simplest and safest way to go about it. Details are given for planning cable runs and choosing the correct weatherproof fittings — since all materials used outside must be stronger and corrosion resistant. Information is also given for low-voltage lighting. Operating at only 12 volts, it is easy to install and does not present the dangers of 120 volts. Like any other exterior lighting, it can make all the difference to the security and attractiveness of your home.

EXTERIOR LIGHTING

A light that illuminates the front and/or back entrance of your home is not only welcoming to visitors it also helps them to identify the house after dark. And most importantly, it also enables you to inspect unexpected callers before you open the door. After all, darkness is on the potential burglar's side, providing plenty of shadows near the house for him to work in. Fitting porch lights and wall lights around the house and lights in the garden will prevent anyone from lurking unseen. Such lighting should be left on throughout the night and can be controlled automatically by a photocell switch which reacts to dusk falling.

For these lights, fit only those that are designed for the purpose. The actual fitting should be weatherproof and the lamp or bulb itself should be held in a moisture-proof cup that surrounds the electrical connections. If possible, try to position a porch light so that the supply cable can be run straight through a wall or ceiling of the porch and into the back of the fitting. If you do have to run cable along an outside wall it should be protected by passing through a length of metal or plastic conduit. A porch light can be installed in the same way as for adding a new light, by taking power from the nearest ceiling junction box, see page 33.

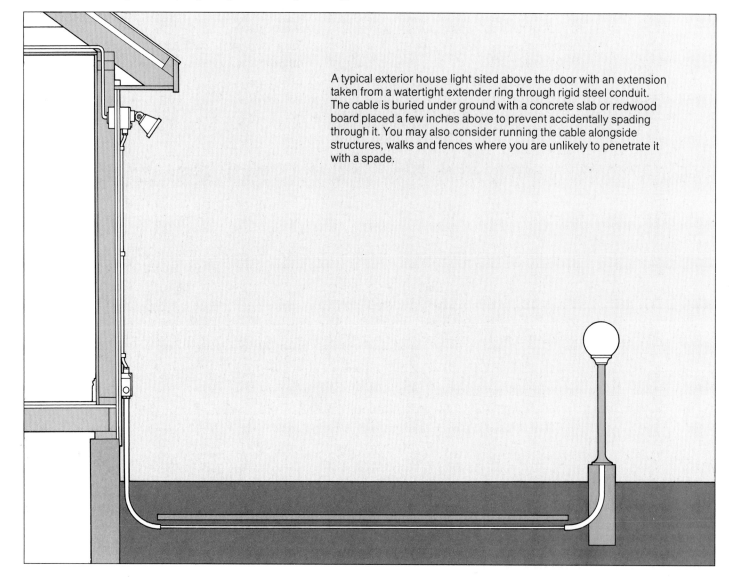

A typical exterior house light sited above the door with an extension taken from a watertight extender ring through rigid steel conduit. The cable is buried under ground with a concrete slab or redwood board placed a few inches above to prevent accidentally spading through it. You may also consider running the cable alongside structures, walks and fences where you are unlikely to penetrate it with a spade.

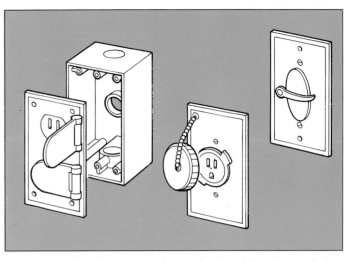

TAKING POWER OUTDOORS

More and more people expect these days to be able
to make nearly as much use of electricity outside the
house as within it – to have a power supply in the
garage, garden shed, summerhouse or greenhouse,
to be able to use the growing range of electrical
gardening tools, to run a fountain or waterfall in the
garden pond, or to provide garden/pool lighting. Of
course it is very tempting to supply all these needs
simply by plugging an extension cord into an outlet
indoors and trailing it artistically across the lawn. The
trouble is that such a set-up is safe only for temporary
use – powering a hedgetrimmer or lawnmower, for
example – and other temporary arrangements have a
habit of very easily becoming permanent (and potentially dangerous). What is needed in all these cases is
a proper permanent power supply.

Assessing your requirements
Once you have decided that you would like to install a
power supply of some kind outside, it is best to work
out beforehand exactly what you expect it to do. For
example, you may want several outlets and good
working light in a garage or garden shed, an array of
fixed garden lighting, or a supply to a pond at the
bottom of the garden; all of these must have new
circuits, fed individually from the service panel in the
house. You may on the other hand want just an
outdoor outlet or a patio light on the back wall of the
house, and these can usually be supplied by simple
extensions of the existing house wiring – no need for
extra circuits in this case.

Planning cable runs
Underground cables should be buried at least 20in
below ground level – and deeper where they cross
areas you dig over regularly, but they can be down as
little as 6in if it is protected beneath a concrete slab.
Alternatively, if the cable is protected by conduit it
can be laid 12in deep.

Check your local code to determine the required
method in your area (conduit is usually preferred).
You may use UF cable which is covered by a heavy
plastic cover if conduit is not used. Use type TW in
conduit. You can of course fix the cable to walls –
boundary or building – but *not* to fences, which could
be blown down and cause the cable to rupture.

Choosing accessories
Whatever type of outdoor power supply you are
planning, it is important to choose the correct type of

Weatherproof outdoor accessories: outlet box with spring-loaded
lids (left), outlet box with screw cap (middle), and weatherproof
switch box (right).

fittings and accessories as well as to install the cable
correctly. Obviously any light fittings must be designed for outdoor use. Outlet boxes out of doors
must be of the weather-proof type, while receptacles
and switches within auxiliary buildings need to be
tough enough to withstand occasional knocks. The
code now requires that all outdoor electrical outlets
must be protected by a Ground Fault Interupter
(GFI) device.

Making the connections
With the cable run complete, you can turn your
attention to the ends of the run. If the cable is run to
an auxiliary building with its own sub-circuits, you
must install a sub-panel there with disconnect, and
connect the circuit cable to its feed terminals.

Within the house the circuit must start either at a
spare fuseway or breaker in the existing service panel
or at a separate sub-panel.

A ground fault circuit interrupter outlet and box for outdoor use.

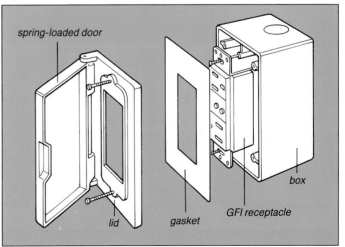

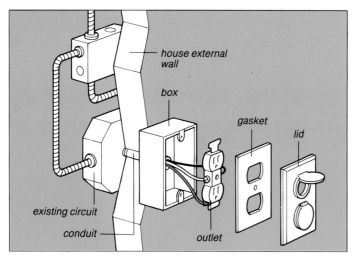

A simple way to bring power outdoors from an existing junction box; running connecting cable through conduit.

provides protection for the appliance when it is plugged into the receptacle. The high-sensitivity type with a cut-off time of 30 milli-seconds is recommended and should be selected.

Low-voltage cable

One or two outdoor lights can completely transform your patio or barbecue area, as well as provide functional lighting for pathways and steps. If you want an outdoor power supply simply to provide decorative lighting or to run a pump for a garden fountain, it is much simpler to opt for low-voltage equipment run from a transformer. Unless otherwise stated by the makers, low-voltage cable can be run on or near the surface of the ground without further protection, but it is as well to inspect it regularly, and do not allow it to trail over stone steps or other sharp edges that may damage the PVC insulation should it be trodden on. The transformer is simply plugged into a convenient outlet in the house or in an outbuilding with a power supply.

For extra protection when using electricity out of doors you must install a GFI circuit breaker at some point. You can either fit one in the sub-panel inside the house or auxiliary building, or else use an outlet incorporating a GFI within it. The former arrangement has the advantage in that the whole circuit is protected; whereas a GFI outlet receptacle only

Outdoor lights controlled with an automatic time-switch which may be by-passed by a manual switch. Plastic-sheathed cable is used for underground installation.

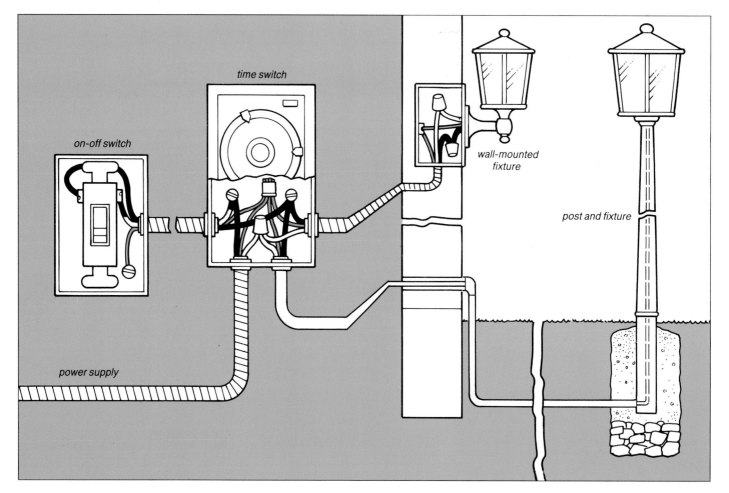

Voltage drop

On long outdoor runs the voltage drop may exceed the maximum permissible figure of 6V – it all depends on the length of the run and the amount of current carried. Make sure that you keep your cable runs as short as possible – less than 75 feet if you are using number 12 wire. Longer runs will mean that you will have to use a larger size wire.

Taking power outside from a junction box inside the house

With an underground supply, the cable will usually be run in conduit throughout the outdoor section. The first job is therefore to mark out and excavate the trench to take it. This should be laid out in straight runs; gentle elbows on the conduit will take care of abrupt changes of direction – for example, where the cable run leaves the house and heads underground.

With the run excavated to the correct depth, lay out the cable along it so that you can measure the precise length required to reach the end of the circuit. Add an extra 3ft to allow for any slight errors and to enable the final connections to be made easily. If you are using plastic conduit start threading lengths of conduit onto the extended cable – this is easier than trying to thread the cable through each length of conduit already in position in the trench. Work from each end of the run back towards and middle, butting each length of conduit up against the next and adding elbows where necessary. When you have added the last complete length of circuit, work along the run bonding the lengths together with solvent-weld cement and lay the conduit in the bottom of the trench. At each end of the run, use elbows and short lengths of conduit to make up the above-ground sections where the run enters the house or outbuilding. Outlet boxes must be at least 12in above the ground. Anchor the conduit securely into the ground by lowering a concrete or cinder block over the conduit before installing the outlet box. Finish off by filling in the trench, covering the conduit in vulnerable areas with pieces of paving slab or roofing tile for extra protection.

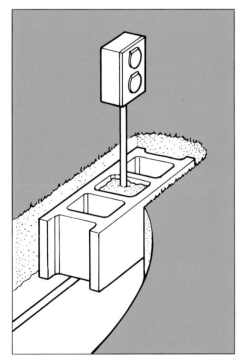

Set conduit to outdoor electric boxes in a concrete block, filling in the core with gravel.

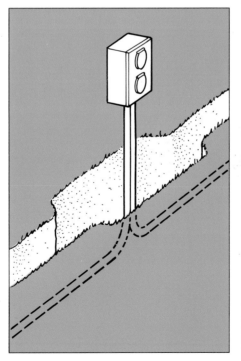

A middle-of-the-run outdoor electrical box with two conduit openings.

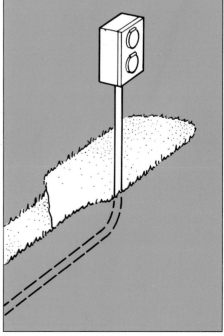

An end-of-the-run outdoor electrical box.

TROUBLESHOOTING

When electricity ceases to flow and an appliance stops or a light will not light, finding out what is wrong and putting it right is usually a matter of some logical detective work. As long as you have a reasonable working knowledge of your electrical system and you are prepared to trace faults systematically, it will only be a matter of time before you find out what is wrong. That is usually the hardest part; once you have found the fault a simple repair is often all that is needed to get the current flowing again.

Whatever the fault, always remember the golden rule and turn off the power before you start your investigations, whether it is at a light switch (to replace a failed bulb) or at the house main disconnect (to investigate an apparent circuit fault). Similarly, always unplug an electrical appliance from the main supply before working on it for absolute safety.

Here is a guide for tracking down the five major types of electrical fault you are likely to experience in your home.

A pendant light fails to work

1 The first thing to check is the lamp. Turn off the light switch controlling the light, remove the lamp and replace it with a new one – or one from another light that you know works if you have not got a spare to hand. Switch on again; if the lamp fails to light, check point 2.

2 The next thing to suspect is a faulty connection between the pendant light cord and the main lighting circuit. To check this you must turn off at the breaker panel, remove the lighting circuit fuseholder from its fuseway or fuse from its fuseway or turn off the lighting circuit breaker; you can then turn the main switch on again to restore power to the other circuits in the house. You may find when you go to do this that the circuit fuse has blown or the breaker has tripped off; this is a sign of a short circuit somewhere on the light circuit – probably on the cord, which is prone to chafing and wear because it can move about in the breeze, and which may disconnect itself from the lampholder terminals as a result. Such a fault will stop the light from working; if in addition a loose wire touches another one, the resulting short circuit will blow the fuse or trip off the breaker.

Unscrew the lampholder first, and check that the wires are securely connected to their terminals. Remake any connections that are lose or broken, stripping off a little more insulation if necessary first before replacing the cover.

Next, unscrew the ceiling canopy and check the cord connections there too, again remaking any that are loose or broken and checking that the wires are firmly attached in the terminals. Replace the canopy.

Now restore the power to the circuit by replacing the fuse or switching on the breaker. If the fuse blows again, check point 3.

3 The last fault you should check for on the pendant light itself is cord continuity; a worn or physically overloaded cord may have a fractured wire concealed within apparently sound insulation. Again working with the power to the circuit off at the panel, disconnect the flex from the junction box and lampholder and test each wire in turn first with a voltage tester and then with a continuity tester. If any wire fails to light the test lamp, replace the cord. If the circuit fuse still blows when you restore the power, there is a fault elsewhere on the circuit (see Whole circuit is dead).

An appliance fails to work

1 The first thing to check is that power is reaching the outlet you are using. Plug in another appliance you know works at the same outlet; if this fails to work here, the fault is at the outlet or on the circuit supplying it (see Whole circuit appears to be dead). If it does work, check point 2.

2 Check the cord fuse on the first appliance; replace it with another of the correct rating for the appliance and test it by plugging it in again. Make sure the replacement fuse is sound by checking it with a continuity tester.

3 Open up the plug and check all the cord connections, remaking any that are broken or loose by cutting the wires back slightly to expose fresh conductors and stripping away a little extra insulation to allow the new connections to be made. Replace the cover, and test the appliance once more.

4 Next, carry out a similar check at the point where the cord is connected to the appliance itself. Unplug it first, then open up the appliance so you can gain access to the terminal block. Remake any connections if necessary, as described in point 3.

5 Check the continuity of each core in the appliance power cord as described in point 3 under light faults, and replace the flex with new cord of the appropriate type and rating if you find any defects.

6 If these checks fail to restore power, then there is a fault on the appliance that should be professionally repaired unless you are qualified to track it down and fix it yourself and also that you can obtain the necessary spare parts.

A whole circuit appears to be "dead"

1 If the circuit fuse has blown or the breaker has tripped off, start by switching off all lights or disconnecting all appliances connected to the circuit concerned. Then replace the fuse with the correct rating, fit a new cartridge fuse of the right size or reset the breaker, and go round the circuit switching on lights or plugging in appliances one by one. If the circuit goes dead at any point, note which light/appliance caused the fuse to blow and isolate it for repairs. Finally replace the fuse/reset the breaker again to restore the power to the circuit.

2 If the circuit is still dead, check all the wiring equipment on the circuit for signs of physical damage or faulty connections where the cable wires are linked to the equipment. Work with the circuit shut off, of course, opening up each accessory in turn to check the state of the wiring at each one; remake any connections that are loose or broken, and replace any part that is damaged.

3 If you know you have pierced a circuit cable during other do-it-yourself work, expose the damaged section and either replace it completely or cut the ends cleanly and reconnect them in a proper enclosure – a junction box in under-floor voids, a conduit box recessed into the plaster for buried cable. Remember to isolate the circuit at the mains before attempting this.

4 If the fault persists, call in a qualified electrician to check the circuit out thoroughly with professional fault-finding equipment.

The whole house system is "dead"

1 If the whole house system appears to have no power, the first thing to check is whether there is a power cut in your area. Remember that with three-phase supply, a fault on one phase only at your local substation will black out roughly half the circuits on the panel, and depending on the order in which the houses are connected to the three phases you may be without power while the lights are still burning at your next-door neighbors.

Report a power cut to your local 24-hour emergency phone as soon as possible.

2 If you have an earth leakage circuit breaker attached to your system, check whether it has tripped off and cut off the supply. If it has, reset it to on – it may have switched off because of what is called "nuisance tripping". If you cannot reset it, this indicates that there is an earth fault somewhere on the system. Tracking this down could be difficult, and it is probably best to call in a qualified electrician to trace it for you.

3 The power may have been cut off by the main service fuse blowing because of a major overload or other fault on the system. If you suspect this has happened, call a licensed electrician to check it and replace it for you if necessary.

MAINTENANCE

As with all household maintenance, it is good practice to make regular checks. And in order to do a thorough job, it makes sense to draw up a list of all the things that should be checked regularly (perhaps the vacuum cleaner every few weeks, for example), as well as those things that should be checked every half year, and others that should be checked once a year. In this way you will establish a good working routine, and not leave anything out which may otherwise become dangerous to you and other members of your household.

• Once a year, check the plugs on all your electrical appliances to make sure they are in good working order. Remember that terminals can become loose, which may cause sparking and start a fire, and where there is water, in a kitchen or laundry room, they would be particularly hazardous.

• Remove the plug cover and check that all the conductors are firmly and correctly secured in the terminals and that there are no frayed wires protruding from the connection.

• Make sure the insulation is sound and that the outer insulation on the cable is firmly fixed in the clamp at the entry point of the plug.

• On all appliances, check the entire length of the cable, and replace any that show signs of wear. Electric iron cables are particularly prone to this.

• Pendant light cables should be checked and replaced from time to time. They may become cracked and brittle because of the build up of heat generated by the light bulb.

• For maximum efficiency, extractor fans should be cleaned regularly and oiled according to the manufacturer's instructions.

If you are doubtful about any part of your electrical system, or in your own ability to complete a repair or installation, then get the help of a licensed electrician who will advise you.

INDEX

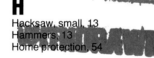